To Rob[...] [...]eline,
with Con[...] [...]e
( )

SPRY

Copy of old painting of Spry sailing in the Bristol Channel. Painted *circa* 1908. Artist unknown.
Reproduced by courtesy of Gloucester Folk Museum, on behalf of painting owner, Mr. Joy.

The Rebuilding of the Severn Trow, *Spry*

AN ACCOUNT BY

C. S. JOHNSON, B.SC., C. ENG., F.I.M.

Rebuilding Project Direction:
The Upper Severn Navigation Trust
in conjunction with
THE
IRONBRIDGE
GORGE MUSEUM

Book publication supported by

PPP healthcare group

PPP healthcare group

The PPP healthcare group is the UK's leading personal healthcare company with over 2 million customers. The group's activities include medical insurance (corporate, personal and international); PPP lifetime care for people looking to insure against the need for nursing care; PPP beaumont care, operators of quality nursing homes; Denplan – the UK's biggest dental capitation scheme, and Principal Investment Management Ltd.

First published in 1995 by The Upper Severn Navigation Trust,
The Wharfage, Ironbridge, Telford, Shropshire TF8 7AW

ISBN 0-9526137-0-0

Typeset from author's disk by Carnegie Publishing, 18 Maynard St, Preston
Printed by Cambridge University Press

Acknowledgements

I wish to express my grateful thanks to those who have helped me to obtain information for this account, from their records and personal recollections.

I especially thank Brian Waterson, former Works Engineer of the Ironbridge Gorge Museum Trust and Archibald Shaw, Consulting Architect, for providing the information for Chapter 3, on the recovery of *Spry* from Diglis Basin in Worcester.

Above all I am grateful to Master Shipwright Alan Williams, for his patience in recounting details of the work involved after the recovery, through the planning and preparations, to obtaining the timber and the actual rebuilding on Blists Hill at Ironbridge. Also for his willing assistance, frequently required, in facilitating my photography of the work in progress. I would also like here to express my personal admiration of the skill and hard work of Alan and his principal assistant, Paul Turner, in carrying out the rebuilding to such a high standard.

Thanks too to the Directors of the USNT, as listed overleaf, for their encouragement, especially former Chairman, David Lawrance, present Chairman Jim Morgan and Major Mark Harwood-Little, former Director of the Ironbridge Gorge Museum Development Trust and Joint Secretary with Mr. Stuart Smith, of the Upper Severn Navigation Trust in its early days, who was particularly helpful in putting me on the track of information on the project's initial development. Also to PPP for their generous support in financing the publication of this account.

Stan Johnson
Wolverhampton.
October 1994.

Upper Severn Navigation Trust Ltd.
Directors (1994)

Contents

List of Illustrations

CHAPTER I

The Organisational Beginnings

The Upper Severn Navigation Trust Ltd

The origin of the restoration, as the work was first envisaged, of the last surviving Severn trow *Spry*, at Ironbridge, is in the setting up of a specialist Trust to recall and even recreate features of the great historical significance of navigation on the River Severn.

In 1976 such a Trust was proposed by the Telford Development Corporation as a completely independent body to the recently formed Ironbridge Gorge Museum Trust (IGMT) and its supporting Ironbridge Gorge Museum Development Trust (IGMDT), with their objectives of illustrating the industrial heritage of the Severn Gorge, 'Birthplace of Industry' and of fundraising respectively.

This independent Trust, to be called the Upper Severn Navigation Trust Ltd. (USNT), was established under the direction of the General Manager of the Telford Development Corporation, Mr. Emyr Thomas, with the assistance of Mr. Frederick Clamp, Secretary to the Corporation and of Col. A. P. de T. Daniell, the Development Officer of the Ironbridge Gorge Museum Development Trust. The Constitution and Articles of Association of the USNT Ltd., were drawn up and application made for Registration early in 1977.

The founding subscribers to the Trust were:

Telford Development Corporation	(rep. Mr. Emyr Thomas)
Salop County Council	(rep. Sir John Corbett)
Wrekin District Council	(rep. Councillor D. C. B. Adams)
Severn Trent Water Authority	(rep. Sir William Dugdale)

Other individual founding members were:

Edmund Bruce Ball Esq., C.B.E.	IGMT
Michael S. Darby Esq.	IGMT
Mark Pearce Esq.	IGM Development Trust
Dr. George F. Williams, J.P.	former MD, Coalbrookdale Co.
Sir George Labouchere	Landowner
Frederick Clamp Esq.	(Telford Development Corporation Solicitor)

Before the Trust was officially established, the Telford Development Corporation, in a letter from its Chairman Lord Northfield, invited a further twenty prominent people to join in support of the proposed Trust, including the then Director of the IGMT. Dr. Neil Cossons, who readily accepted, giving the Museum Trust a further, significant, representation. The other people accepting the invitation were:

Sir Frank Price Esq.	British Waterways Board
Basil Greenhill, Director	Maritime Trust
Malcolm Braine Esq.	Boatbuilder
Col. A. P. Sykes Esq.	Landowner
Jack Thompson Esq.	Landowner
H. Brook Weld Forester	Landowner
J. Benthall Beard Esq.	International Insurance
David Smith Esq.	Midland News Association

After a prolonged period, the USNT Ltd. was registered as a Company in 1977, but it was not until the 22nd September 1978 that it was granted Charitable Trust status. The delays were due to many reasons but it is recorded that it took three months to obtain the signatures of the seven founding members!

The original logo on the newly formed USNT Ltd., letter head was of a line drawing of a bow-on view of a square rigged Trow under sail. This was changed at an early date to a side elevation line drawing of a sloop rigged Trow without sail.

Major Mark Harwood-Little, who was appointed Director of the IGMDT in 1978 in succession to Col. Daniell, and Mr. Stuart Smith, who was appointed Director of the IGMT in succession to Dr. Neil Cossons, about the same time, took over their predecessors' keen interest and involvement with the new Trust as its Joint Secretaries.

Memorandum of Association

It was considered advantageous that the USNT Ltd. should be a completely independent body to the Ironbridge Gorge Museum Trusts. The objectives and enabling powers as set out in the Memorandum of Association of the USNT Ltd. were therefore very comprehensive, as can be seen in the the Appendix, where they are reproduced almost in full. In summary the Trust was authorised to follow and employ the following Objects and Powers:

Objects

For public benefit and education:

1) To acquire, equip, preserve in sailing condition, one or more Severn trows and such other craft as were operated on the Upper Severn or its associated waterways.

2) To operate and exhibit them and to demonstrate their historical significance.

Powers

1) To buy and sell, lend, exchange, mortgage.
2) To build, alter etc. waterway buildings and land.
3) To arrange for public viewing.
4) To carry out research.
5) To raise funds, to borrow and invest money.
6) To employ staff.

As can be seen, these cater for all aspects of investigatory and practical work, fund raising and display, concerned with the navigational history of the Upper Severn and its contributory waterways.

There were some misgivings expressed at the time of formation, concerning the interaction of the new Trust with the existing Museum Trusts, particularly by founding member Mr. Mark Pearce, an IGMDT trustee, but his views were not shared by the majority. In practice it is apparent that these powers were not fully exploited and no doubt for simplification of administrative control of both practical work and fundraising, the USNT activities tended to be affiliated to the other two Trusts. Initially this worked well, with good cooperation and results obviously facilitated by the close associations arising from the Joint Secretaries being the Chief Executives of the IGMT and the IGMDT. Due to the closeness of the association, over a period of years the accountabilities seemed to become somewhat blurred and the USNT Board of Management became more of a 'staff' advisory body to the executive activities of the other Trusts, despite its authority to operate with complete autonomy in persuing its objectives. Later, under the financial pressures of a recession in the early 1990s, this loss of independence caused some difficulties in agreement in persuing USNT priority objectives.

Early Meetings

Shortly after the registration of the USNT Ltd. as a Company and a Charity and before its first official meeting to elect a Council of Management, which took a year to arrange, an unofficial meeting took place on the 11th December 1978 at Meachams boatyard, in Pelsall, near Walsall, at the invitation of Mr. Malcolm Braine. It is of interest to note that the Beard family of Coleford Old Hall, of which Malcolm Braine was a descendant, had been associated with trow building and ownership for at least 200 years and in fact owned the last trow to trade on the River Severn, the 'William' of Broseley. The meeting was called to discuss the possibility of building a trow in Ironbridge instead of commissioning a replica to be built elsewhere. The size of the trow proposed was one of about 65 feet overall length and 14 feet beam which would be representative of an upper Severn trow of the mid-eighteenth century. The site of the old Bower Shipyard on the south bank of the Severn in Ironbridge, opposite the Severn Warehouse, was also proposed. It was subsequently established that the site belonged to the Telford Development Corporation and being contacted, they had no objections to the site being leased for this purpose, subject to the usual planning approvals and no interference with already leased fishing rights. This preliminary discussion and investigation was felt to be encouraging to the aims of the USNT.

Present at this informal meeting were Messrs. M. E. Braine, R. Barker (Marine historian) and Major Mark Harwood-Little. An apology for absence was recorded for Mr. Stuart Smith.

It is interesting to note in the meeting record that the *Spry* was mentioned as being a down river craft and too big for inclusion in a USNT project. Also that *Spry* was not, strictly speaking, a typical trow, as her planking was carvel fashion throughout, instead of being carvel bottomed and clinker sided as trows usually were.

Following this meeting, in March 1979, a very comprehensive paper was prepared by Mr. Richard Barker, a knowledgeable marine historian, giving a detailed consideration in favour of the proposal discussed, to build a replica trow on the Bower Shipyard site and display it afloat on the river in Ironbridge. The paper included an appreciation of some of the problems involved, and also an outline estimate of cost, to assist further discussion on a formal basis, when the USNT Council of Management was elected.

Eventually, on the 17th October 1979 the inaugural AGM of the USNT took place and it was followed by an open meeting in which, amongst other interested parties, the local fishing club were present, concerned to be assured their activities would not be detrimentally affected. This was followed on the 12th November 1979 by the first meeting of a Council of Management elected at the AGM. The Council resolved to create two working Groups. The first to be concerned with the display of a trow or similar vessel in the Gorge and the second to be concerned with investigating and recounting the history of the upper River Severn, this being defined as being from Stourport and Pool Quay in Welshpool.

It is from the first named that the events recorded in this book derived.

The Severn Trow Preservation Trust

At about the same time as the formation of an Upper Severn Navigation Trust was being initiated in Ironbridge, on the 11th April 1978 another body was formed specifically to try to save the lower River Severn trow

Spry which had been lying in the mud in Diglis Canal Basin at Worcester for many years. She was recognised as being the only remaining trow and therefore of considerable historic significance. She appeared to be substantially intact and therefore salvageable. The subscribers to this Trust, STPT, were Peter Dennis Hingley of Stourbridge, a librarian by profession and John Edwards of Worcester, a photographer.

A year later, under pressure from the British Waterways Board, which wanted to use the berthing space occupied by the *Spry*, the Severn Trow preservation Trust found itself unable to raise sufficient funds to proceed with the salvage. In April 1979 the Trust published an S.O.S. for funding support in the Press which caught the eye of the promoters of the USNT and on the 27th April 1979 Mr. Stuart Smith wrote to Mr. Peter D. Hingley of the STPT enquiring for the details of their position. This was the first contact with the *Spry* by the USNT and while current thinking there was still, at that time, that the *Spry* was not a suitable subject for the USNT to be involved with, this judgement was to change in due course.

The basis for future activities

At the end of 1979 therefore the scene was set, so to speak, for the events which subsequently followed. The USNT Ltd. was formed with clearly defined objects and powers, to persue illustrating the history of the upper Severn and with an early key objective of building a replica trow on the Bower Shipyard site and displaying it on the River in Ironbridge. The Severn Trow Preservation Trust, formed to preserve the trow *Spry*, was unable to progress in its objective due to lack of financial support and there had been a preliminary contact between the two groups.

Before proceeding with the story of the events which followed, in the next Chapter, the author would like to outline the historical background which makes the ultimate outcome so significant.

CHAPTER 2

History: The River Severn and its navigation, the significance of trows and of *Spry* particularly

The beginnings

The Geological History of the River Severn is of great significance to later developments in its navigation and the importance of trows in that activity.

Millions of years ago decaying vegetation and sedimentary materials were deposited across the region while it was covered by shallow waters, which eventually became layers of coal, associated with ironstone and clay, overlaying earlier depositions of limestone. In the course of further evolution and convolutions of the Earth's crust, these layers were raised and formed a continuous elevated ridge of land from Wenlock to the Pennines, as they are now called.

The Severn Gorge was formed through this at the end of the last Ice Age, some 60,000 years ago, by the joining of two separate rivers, one rising in the North in the Welsh hills, flowing initially South West along the course of the present upper Severn but then flowing North to empty into the Dee Estuary. The other, rising near where Wolverhampton is now, flowed South, following the course of the present lower Severn, emptying into the sea through what is now the Bristol Channel.

As the ice retreated, the Northerly drainage into the Dee was blocked causing a vast lake to form behind the Wenlock/Pennine Ridge which eventually overflowed to the South. The scouring action of the outflowing waters caused the rapid erosion of the Gorge through the Ridge and permanently diverted the flow of the northern river to combine with that of the southern river, to form the present river Severn. In the scouring action the underlying strata of coal, ironstone, clay and limestone were exposed for easy access by later human inhabitants. Thus the 'scene was set' for the momentous developments which made the Severn Gorge the Worlds 'Cradle of Industry' in more recent history, in the 17th and 18th Centuries.

The history of River Severn navigation

The river has most likely been used for navigation and the transport of goods from very early times and inferences from the siting of Roman settlements near its banks, including Viroconium in what is now Shropshire, strongly suggest that the river, including its upper reaches, was used by them for transportation.

Apart from the recorded use by the Vikings for forays upriver as far as Gloucester, more complete records from the thirteenth century show that river navigation was well established by then, initially for lead from mines near its upper reaches, also for rafting timber downstream. Of particular interest concerning the subject of this book, is a record in the Shrewsbury Corporation accounts for the Winter of 1269–1270 of the transportation of paving slabs from 'Tewsburia' – the *Spry* having been built 6 centuries later for the carrying of stone!

The name 'trow' (pronounced as rhyming with 'crow') seems to have been derived from 'trough' or 'trog' which have a similar connotation of being hollow wooden vessels, originally hollowed out logs, for holding water, excepting in the case of the boat the water is outside! This name was specifically used for vessels on the River Severn and also for boats used for the much smaller traffic on the River Tyne, although boats of generally similar design evolved on other rivers. The first recorded use of the name seems to have been in 1411, when trow owners in the Bewdeley area formed an association which sought by violent means, to prevent other boat and raft owners from using the River and a petition was presented to Parliament to counter this and re-establish the right of free navigation for all.

These were violent times and there are also records of piracy of the trows, for their valuable cargoes, as they passed through Gloucestershire, by the inhabitants of the region. Lawlessness and violence continued to be associated with trow crews into much later years as they were apparently notorious for acts of vandalism and poaching, especially when the vessels were moored for periods awaiting suitable river conditions to continue the voyage.

In the sixteenth century a contemporary traveller, Leland, recorded that at Bewdeley bridge he saw '*many flatt vessels carry up and down all manner of merchandices*'. Other records also show that the main cargoes at the time were local products such as cloth, flannel, wool, hides, cheese, butter and grain and the goods for which they were exchanged, which included hops, spices, wines, and brandy, and other imports, brought to Bristol by seagoing ships.

Towards the end of the sixteenth century the transport of coal from the upper reaches began for domestic use and rapidly became one of

Figure 1: Photograph of model of River Severn Gorge in Ironbridge, showing activities in 1796.

the most important cargoes as supplies of timber dwindled and urban living developed further.

In the eighteenth century, with the coming of the Industrial Revolution, the traffic on the River increased greatly, reaching its peak in the late 1700s. During this time, for almost a century, the Severn became the second busiest river in Europe, after the Meuse.

G. Farr, in his most valuable paper on his researches into the commercial history of the River, reproduces a letter which appeared in the *Gentleman's Magazine* in 1758, written by Mr. G. Perry of Coalbrookdale and dated 15th June 1758, which was a detailed account of the intense activity on the River at that time and which is felt to be of interest to reproduce here again in full:

> *This river is of great importance on account of its trade, being navigable by vessels of large burthen, more than one hundred and sixty miles from the sea without the assistance of any lock. Upwards of 100,000 tons of coal are annually shipped from the collieries about Madeley and Broseley, to the towns or cities situated on its banks, and from these into adjacent counties also, great quantities of grain, pig and bar iron, manufactures and earthenwares, as well as wool, hops, cider and provisions, are constantly exported to Bristol and other places, from whence merchant goods are brought by return. The freight from Shrewsbury to Bristol is about 10s. per ton and from Bristol to Shrewsbury 15s. The rates to intermediate towns being in proportion.*

This traffic is carried out with vessels of two sorts, the lesser kind are called barges and frigates, being from 40 to 60 feet in length, having a single mast, square sail and carry from 20 to 40 tons, and trows or larger vessels which are from 40 to 80 tons burthen; these have a main mast and top mast, about 80 feet high with square sails, and some having mizzen masts; they are generally from 16 to 20 feet wide and 60 feet in length, being when new and completely rigged, worth about 300L.

This number being greatly increased of late, caused in May 1756, an exact list to be taken of all the barges and trows upon the River Severn, whereby the increase, or diminution of the trade may be estimated in future times, which were as follows:

Name of Port	*Owners*	*Vessels*
Welchpoole and Poolestake	4	7
Shrewsbury	10	19
Cound and Buildwas	3	7
Madeleywood	21	39
Benthall	8	13
Broseley	55	87
Bridgnorth	47	75
Between it and Bewdeley	8	10
Bewdeley	18	47
Between it and Worcester	7	13
Worcester	6	21
Between it and Upton	2	2
Upton	5	5
Tewkesbury	8	18
Evesham-on-Avon	1	2
The Hawe	3	4
Gloucester	4	7
Totals	210	376

Since taking the above list, the number is advanced to near 400 vessels. They are navigated by 3 or 4 men, who, being robust and resolute, may be esteemed a valuable nursery of seamen.

An excellent impression of this volume of traffic on the Severn, still flourishing 20 years later than this account, in 1796, is today to be seen in the remarkably detailed historic model of the stretch of the River from Broseley and Benthall, by Ironbridge, named after the still famous landmark built there in 1759, and down to Coalport, which is exhibited in the Ironbridge Gorge Museum Trusts 'Museum of the River' on the river bank at Ironbridge. (Figure 1)

This represented the peak period of activity in the river and from the early part of the nineteenth century, competition from the canals and their reliability for navigation, caused a rapid dwindling of the traffic on the river, firstly on its higher reaches. This was accentuated by the advent and rapid development of the railways from the middle of the nineteenth century. The use of the smaller upriver vessels virtually died out in the 1820s, although the last boat built in Shropshire was launched in 1868 and in 1895, the last trading barge coming down from Ironbridge, had a tragic end by being wrecked on striking the bridge at Bridgnorth.

As a last vestige of activity on the Upper Severn, in 1899, a barge was towed upriver from Gloucester to Shrewsbury, to prove the rights of the Towing Path Company. After 1909 there was no commercial navigation at all above Stourport.

Trows continued to be commercially viable much longer in the lower reaches of the Severn, particularly navigating across the river estuary, and some continued in service until into the beginning of the twentieth century. *Spry* was one of these vessels. They plied between the estuary ports where land connections tended to be lengthy and relatively costly, the lowest river crossing being at Gloucester. The shipping of stone was a common activity and many of the trows for this trade were centred on the River Avon. A number of these were actually built at Honey Street, in Wiltshire and on the River Parrot not far from Bridgwater, in addition to yards remaining on the lower Severn.

Trow Design

As described in Mr. G. Perry's letter, quoted above, trows were the larger sailing vessels plying the river and these evolved into two types. The original trows evolved for upriver use, sailing down as far as Gloucester interacting with small coastal sailing ships called 'cogs' for goods coming from or going to maritime ports. It seems likely as trading opportunities developed in the Severn estuary, that trows were modified and enlarged to be capable of sailing further down the tidal reaches of the estuary to specialise in the traffic in this area. Others became to be built larger and with greater draft specially for this purpose. This was particularly the case in the nineteenth century when competition from the canals and later railways initiated a decline in upriver traffic. A unique example of the former type of modified trow, is one of which the remains were found embedded in the river bank near Lydney and which was the subject of an archaeological examination in 1992. (Ref. Ironbridge Gorge Museum Trust Archaeological Series No. 27.) Some of the specially built estuary trows were built upriver, as high up as

Ironbridge, but were floated downstream to be fitted out downriver to ply the lower reaches, below Worcester and in the Severn estuary, and were too large to return upriver. The *Spry* is this type of estuary trow, but was built on the estuary, at Chepstow.

Trows were characterised by having an open hold between small fore and after decks without even a minimum of side decking. They were very full in the bows though with quite a fine run aft. They were flat bottomed, round bilged and shallow draughted, to best navigate the river shallows and to be able to reach as high as possible upriver. Information on the details of design and rigging tends to be somewhat unreliable, as it derives almost entirely from paintings made by artists who are understandably likely to have had little nautical knowledge and to have been primarily interested in pictorial effects. However it seems likely that trows were originally double ended with the squared transom stern developing later. The hulls were stoutly built of oak framing throughout, onto which a flat carvel bottom, planked in Scots pine, abutted an oak keel plank and extended out to just above the turn of the bilges, where oak clinker planking made up the bow, stern and topsides. Later carvel planking became to be used throughout. The early versions appear to have been rigged with a square main and topsail on the main mast and had a mizzen mast rigged with a lateen sail.

The relatively shallow draught and flat bottom were also a considerable asset for the later development into larger vessels for estuary navigation where, with the considerable tides and shallows, the shallow draft and flat bottom gave the vessels the ability to take the ground to discharge cargo where no port facilities existed . Trows of this type had fore and aft rigging as cutters, a rig better suited to open water sailing. For the rougher waters of estuary navigation, the open hold was protected by side cloths to prevent waves being shipped aboard while loaded to the maximum. Even so there was a statutory limit to how far out towards the open sea these vessels were permitted to sail because of the risk of swamping in rough weather.

The history of the *Spry*

Spry was one of the last trows to be built. She was an addition to the small fleet of sailing vessels, which included the trows *Eliza* and *Nelly*, owned and operated by a Chepstow stone merchant, William Davies. The stone being limestone from riverside quarries near Chepstow, probably at Lancaut, situated upriver towards Tintern. *Spry* was built by William Hurd, whose yard was actually at Chepstow on Gunstone Wharf just below Chepstow bridge and she was launched in 1894. She was registered at Gloucester on October 25th of that year. It was not many years after this that river estuary navigation by such trows died out altogether.

Spry therefore was one of the very last of a long line of wooden sailing ships which served the commercial activities and developing industries along the Severn Valley and adjacent regions for at least six centuries, and is therefore a very important historic link with a bygone era.

She was one of the larger trows to be built at 71 feet 6 inches from forepart of stem to aftside of stern post, main breadth to outside of plank 18 feet 3 inches and with 6 feet 3 inches depth of hold. Her gross tonnage was 41.4 and registered tonnage 36.45.

Spry's Construction

Apart from relatively minor changes to her superstructure (mentioned later in Chapter 5), *Spry*'s basic hull construction was not altered during her working lifetime. She was built on substantial trow lines with a bluff bow, square transom, round bilged and flat bottomed.

She was substantially built, being an estuary trow and for her duty as a stone carrier. The following description is from the examination of the vessel since she was recovered. To help the non technical reader, a glossary of the boatbuilders terminology used is given in Appendix 5.

Her construction consists of an approximately 65-feet long keel, 'sided' (boatbuilders term for width) at 8 inches and 'moulded' (i.e. depth) at 9 inches, in two approximately equal lengths of elm, with a stepped scarf joint. The aft length, while moulded at 9 inches for most of its length, sweeps up over the last 10 feet to be moulded at 12 inches at the stern post. The stem is of oak, sided at 8 inches, with a general moulded depth of 12 inches. It is left square at the stem head but has a full 'bearding' (taper) down to the forefoot, where it tennons into the keel. Immediately behind this is the fore deadwood in oak, sided at 10 inches, curving approximately 30 inches up the stem and extending aft 54 inches over the keel. Sitting above this and immediately backing the stem, is the apron also in oak, again sided at 10 inches and moulded at 6 inches. Aft, the stern post in oak, is sided at 8 inches and moulded at 10 inches. This also tennons into the keel. Backing this and bracing between the keel and stern post, is the deadwood, in oak, sided at 8 inches with approximately 9 feet on each arm, and some 36 inches through the corner, resembling an enormous shelf bracket. The assembly is fastened together with bolts of ⅞-inch round bar, peened over large steel washers inboard.

This assembly of the keel, stem and stern post, with their backing deadwoods, forms the backbone of the vessel and they are rebated to carry the planking. In the stem and stern post the rebates are cut square to the run of the ends of the planking, but these rebate angles change to more acute ones as they approach the keel and in the keel itself. This rebate takes the form of a 45° chevron. This is usual where a vessel is particularly flat bottomed.

Onto this 'back bone' are set up the 'rib cage' of frames to carry the planking. There are 40 frames, plus knight heads, hawse timbers, transom and fashion pieces, all in oak. Starting from for'd, first is the knight head and then the hawse timber, each piece slightly wedge shaped and 4 inches through. Their widths at the head are approximately 10 inches and tapering to just 3 inches at the bottom and about 8 feet long. These stand up 2 feet above the frames, flush with the top of the stem. They are spiked edge-wise, with ¾ inch round dumps up to 30 inches long, first through the hawse timber, then through the knight head into the apron and eventually into the stem. This is the construction both to port and starboard aft from the stem. These timbers take much of the loading from the forestay, bowsprit and also anchor chain when the vessel is anchored. Locking all into the hull, above deck level, they are pierced with holes approximately 7 inches in diameter, which take the cast-iron hawse pipes.

Behind this bow assembly, are the first three frames which are called 'cant frames'. They abutt the fore deadwood, into a small housing, and tend to follow the sweep of the planking, unlike the frames proper in the main part of the hull which run at right angles to the keel (centre line) and passing over it. These three cant frames, on both port and starboard sides, run only slightly canted from the centre line and, as with all *Spry*'s frames, are sided at 4 inches, moulded 6 inches at the keel tapering to 5 inches at the gunwale (gun'l). They are also double.

This 'double sawn' construction, that is the nailing of two 4-inch widths of timber together side by side, with the joints staggered to overlap their neighbours in the main framing, gives a much stronger frame. While there are 40 frames, there are in fact, 80 full girths of framing in the hull. All the frames have limber holes cut in them, about 1½ inches wide and 1 inch high and 9 inches each side of the centreline along the run of the keel, to allow longitudinal draining from each frame section aft, to the bilge pumps. The cant frames also have limber holes in their lowermost corners where they abutt the deadwoods.

Behind the cant frames come the main frames, of which there are 32, at 20-inch centres. Each frame consists of two floor timbers, two futtocks, two second futtocks and two top timbers completing the sweep up to the gun'ls. In all eight pieces, each piece staggered to cover the butt of its neighbour and spiked through.

Frames numbers 9 to 33 inclusive, stand up some 9 inches above the others, which finish just below the top edge of the shear strake which takes the raised hatch coming.

Aft, frames 36 to 40 are again cant frames, abutting the aft deadwood and finally the transom and fashion pieces at the stern.

Along the centreline of the frames, and fastened through to sandwich them between it and the keel, with 38, 1-inch square bolts, is the keelson, probably originally fashioned from pitch pine. This is approximately 60 feet long, sided and moulded 16 inches square. In *Spry* as recovered, this had obviously been replaced at some time, possibly around 1911, from records of the times *Spry* was under repair. This is to be expected from the very hard wear of about 15 years carrying stone.

The top ends of the frames are secured by a 10 × 3-inch section sheerstrake outside the frames and a beam shelf inside the frames, running around the perimeter of the vessel. The beam shelf, also of 10 × 3-inch section, as its name implies, carries the deck beams for'd and aft. The deck beams are sided at 4 inches and moulded at 6 inches with 4½ inches of camber, bearing in mind that the hatch is raised above the level of the deck. There are three very large cross hatch beams sided at 10 inches and moulded at 13 inches, with 2 inches of camber. As these are above the shelf there is nothing substantial upon which to support them, therefore they are each secured by two lodging knees (brackets) and supported from underneath by one hanging knee at each end. These are made of wrought iron, and bolted through the hull. The hatch beams give the necessary rigidity across the top of the hull.

The external planking of pitch pine or elm, of 12 × 2-inch cross section amidships, tapering to 4 inches at the ends of the boat, is of carvel construction throughout. The hull is lined with ceiling planking, also of 12 × 2-inch section pitch pine.

The external planking would have been made up of elm, where totally submerged and pitch pine on the 'wind and water' line and above. The planking consists of 13 strakes per side, including the sheer strake and two 'stealers'. (A strake is a series of planks end to end running the full length of the vessel and a stealer is a series of planks end to end, but finishing before the ends of the boat.) In *Spry* the stealers are in the way of the bilge, one running from for'd, finishing on the aft quarter and the other running from aft finishing on the for'd quarter. The use of elm for submerged planking is because it withstands total submersion very well and also has a good abrasion resistance for grounding. It is not suitable however on the wind and water line. On *Spry* the garboard strake was in elm for its entire length. The second strake was, from for'd, pitch pine-elm-elm. The third strake was, from for'd, pitch pine-elm-pitch pine, as were the rest of the strakes, out to just above the turn of the bilge whence pitch pine took over completely. The bilge keels were entirely of elm.

The ceiling planking was of pitch pine, as were the bilge stringers. The ceiling commenced 4 inches from the keelson, leaving a gap for access for cleaning the limber holes, which was covered by a removable plank, the planking being taken out to the two bilge stringers, thence up to the beam shelf.

The bilge stringers were each made up of a 8 × 5 inches by 38 feet length and a 5 × 5 inches by 42 feet length, side by side, the 8 × 5 inches being above the other. These were bolted through from the outside, through the bilge keels and the outer planking, through the frame and finally through the bilge stringer, where the bolts were peened over a washer. ⅝ inch by 22-inch round bolts were used, two bolts per frame. The bilge stringers ran between frame numbers 8 to 29.

The beam shelf is of 10 × 3-inch pine and runs around the inside of the frames, to just the depth of the deck beams, and below the top of the frames. The beam shelf is for resting the deck beams upon. It is also a major longitudinal structural member, as are the keel and keelson, bilge keels and bilge stringers. Similarly the sheerstrakes are also structural members.

Spry has only a small fore and after deck, the hold being completely open with no side decking at all along the gun'ls. The gun'ls themselves are capped with an 11 × 6 inch baulk of pitch pine, with a 2 inch square rebate its top outside corner to take the canvas side cloth and batten.

15 feet for'd from the transom is decked over and partitioned from the hold to form a master's cabin, reached from the deck by means of a sliding companion hatch and steps down. The headroom is very limited at 4 feet 6 inches under the deck beams, which are sided at 4 inches and moulded at 6 inches, with a 5-inch camber and have ½-inch drip beads moulded on the under surfaces. The deck beam number 6, aft of frame number 33 is the margin of the hold and is sided at 5 inches. Deck beams 7 and 8 carry the carlings for the scuttle to port. Deck beams 9 and 10 carry the carlings for the cabin skylight amidships and the mooring bits port and starboard. (Carlings are the fore and aft frames between the deck beams onto or through which fitting are secured.) On the forward side of deck beam 9, is carved 'Certified to accommodate the master'.

The deck planking is of 5⅞ inch by 3 inch section, originally pitch pine. Around the edge is a cover board made in one width at 10 × 3 inches of oak, pierced to take the stanchions and nailed down into the shear strake outboard and the deck beams inboard. There is no joggled margin piece, the deck planks run from a shaped edge at the stern for'd to a 1-inch overhang past deck beam number 6 at the hold and are finished with a bullnosed end. The plank ends over the transom also overhang by 1 inch and are similarly finished.

Around the deck, the bulwarks are 2 feet high, consisting of 4 planks 6 inches wide. The lower one is 1½ inches thick with a ½-inch bead moulded in its top outboard edge, also a 'cavetto' line moulded in 4 inches above the cover board. This is also moulded through the hatch sides. Traces left show that the line was picked out in gold leaf. The other three planks in the bulwarks are 1 inch thick. These are nailed onto the bulwark stanchions of 4 × 4 inches section at deck level and 4 × 3 inches at rail level, extending down through the cover board some 3 feet and fastened through the external planking, stanchion and ceiling planks with trunnels (wooden nails), two per stanchion, the trunnels being split and wedged outboard. The reason for this archaic form of fastening is because there is a high risk of the bulwarks being damaged with the vessel working in and out of harbours very frequently and it is much easier to back out a trunnel than it would be to remove nails in order to lift out a broken section for repair.

The inside of the Master's cabin is fitted with a partially screened bunk space with a horseshoe shaped settle around the stern and there is wooden panelling around the hull. Adjacent to the bulkhead, the main and auxiliary bilge pump casings rise through the cabin for discharging into the scuppers of the deck above. A raised skylight and a glass deck prism provide some daylight illumination, the former also providing the master with an outlook for'd from below and is the only place below deck, where the Master could stand upright.

16 feet from the bow is similarly decked over and partitioned from the hold, forming a cabin for the crew, with deck beams of the same dimensions as in the stern . The first deck beam for'd of frame number 3 and the second deck beam aft of frame number 4 carry the carlings for the anchor winch legs and also the carlings for the samson (or pawl) post. The anchor winch carlings are of 9 × 3 inches pierced with a 9 × 4 inches square hole to take the leg of the anchor winch. On the aft side of deck beam number 2 is carved 'to accomodate (as spelt) 4 seamen. Between deck beams 3 and 4 are carlings for the for'd scuttle amidships and between numbers 4 and 5 are the carlings to carry the cargo winch legs, made out of two 9 × 3-inches timbers pierced by an 8 × 4-inches square hole for the legs. Deck beam number 5, for'd of frame 9, being the last deck beam before the hold, is sided at 5 inches. The crews cabin was probably only fitted with two bunks, one each to port and starboard, although there is room for superimposing two more bunks. There is a stowage locker right in the fore deck.

Spry's original rigging can only be deduced from the few illustrations of her which exist, notably the oil painting owned by Mr. Joy and exhibited in the Gloucester Folk Museum, (frontispiece), which has obviously been painted and lined by, or in cooperation with, a knowledgeable marine expert, possibly one of her Masters, for the rigging is remarkably detailed. She probably originally had an 80-foot, two piece, main mast which could have been lowered by a winch on the foredeck,

Figure 2: Photograph of half-inch to the foot scale model of *Spry* made by Mr. Arthur Aston.

also doubling as a cargo winch. She would have had a 48-foot boom, 30-foot bowsprit, gaff, staysail boom, cargo boom and a topsail yard. None of these were, of course, still fitted in the hulk. To complete the outline picture of her construction, *Spry* had an anchor winch on the foredeck which was still on board as she was found, although it had collapsed into the cabin space.

The photograph, Figure 2, of an accurate, ½ inch to the foot, scale model of the vessel, made by Mr. Arthur Aston of Coseley, West Midlands, taken from measurements of *Spry* during her rebuilding as described later, gives a very good impression of *Spry* as she would have been built in 1894.

Early voyages

From *Spry*'s original official logs of 1894 – 1912, still in the Gloucester Record Office,* in the first two months after her commissioning on 25 October 1894, she made 12 voyages in the river stone trade between Cardiff and Chepstow. Her master was William Halling of Frampton and her crew a mate and one ordinary seaman and also 13-year-old Harry Halling, as ships boy, probably a relative of the master, under

* Information from extracts dated 14th October 1977 from P. D. Higley of the Severn Trow Preservation Society

whom he had previously served in the *Nelly*. In 1900 she was trading up the River Severn with her more usual crew of three men and in 1902 in another source she is mentioned as bringing a cargo of 100 tons of stone to Gloucester.

In 1908 her master was at first William Wayman of Chepstow, making 10 trips in her first half year between Cardiff, Chepstow, Newport, Gloucester and Bristol, but she was laid up at Chepstow on 1st July to 30 August and thereafter had a new crew commanded by Thomas Aldridge of Framilode. Under him she made 20 similar voyages in the second half year, including to Lydney this time, although she was under repair at Chepstow from the 29th October to 4th December.

In 1910, William Davies died and his small fleet was dispersed. *Spry* passed to George Nurse of Bristol. In 1912, in the last log held in the Gloucester Record Office, she still had a crew of three and while all her voyages are listed, unusually, her cargoes were not recorded. Between 11th July 1912 and 4th September 1912 she made 12 voyages, visiting Charston, Bristol, Wick River, Gloucester and Lydney.

From information given in a letter dated 25th February 1984, by Mr W. J. Joy, whose maternal grandfather, Thomas Aldridge, was master of *Spry* just before the first world war, she made several voyages, not detailed in the log books of that period, to the Chaston sands in the Bristol Channel, to obtain sand for a glass bottle factory in Bristol. The vessel was allowed to settle on the sand bank as the tide went out and the crew shovelled the sand into wheel-barrows which were run up a plank to be tipped into the hold. According to Mr. Joy's information also, Thomas Aldridge was asked to captain *Spry* to convey materials to France because of a shortage of shipping during the war, but refused, principly because by that time *Spry* was in a bad condition, for which same reason he had earlier refused an offer to buy the vessel. (Mr. Joy was also able to give some details of the furnishing of the cabins which was very useful in the rebuilding of the vessel).

In the 1914 Mercantile Navy List (M.N.L.), she was described for the first time as being a ketch, probably having been converted in 1913, and in the M.N.L. for 1916 she was recorded as being 'of Bristol' and as being owned by Alfred J. Smith Ltd., of 9, Queen Square, Bristol. In 1923 she was back in Gloucester, the owners being given as Jacob Rice and Sons Ltd.

At some later date, she was cut down to a 'dumb' barge – i.e. to be without sails or motor and depending on being towed by a tug – like many other remaining trows. The M.N.L. lists her as a ketch until 1936 but two photographs exist of her in her reduced state, which are dated August 1932. It is possible that the M.N.L. information may not be quite accurate. In 1936, her Certificate of British Registry, dated 11 May and issued in Bristol, she was described as 'no rig', so it can only be stated from records available that she was cut down sometime before then.

In 1950 she entered Diglis Basin in Worcester and remained there, being converted into a floating and later sunken workshop base and passing through various ownerships.

Being particularly stoutly built for the heavy duties of carrying stone and due her own particular circumstances, *Spry* alone survived through all these many adaptations and usages until she was discovered in the late 1970s at Worcester, in her ignominious role of semi floating workshop in the mud of the Diglis Canal Basin, by enthusiasts who recognised her historic uniqueness as the last of a very noble line indeed and the saga of her recovery and rebuilding began.

CHAPTER 3

The Recovery

(From discussions with Brian Waterson, Archibald Shaw and file records.)

In the first Chapter, the formation of the Upper Severn Navigation Trust Ltd. (USNT), in Ironbridge, was recounted, together with its objects of recalling and displaying the great historical significance of the upper River and its navigation in the development – and decline – of the 'Birthplace of Industry'. It was also mentioned that another, unrelated Trust, the Severn Trow Preservation Trust (STPT) was formed to prevent a particularly unique survivor of the trow era, the Severn trow *Spry*, a large trow which used to ply the lower River, in its estuary, suffering the same fate as all the other trows and being lost altogether.

The situation therefore, was that during 1980 and early 1981 the USNT were investigating details and funding prospects to build a typical replica upper Severn trow and the STPT, after raising limited funds and carrying out some preservation work on the *Spry*, found, in 1981, that they could not find the resources necessary to carry on and achieve their objective by themselves.

In the previous Chapter the geological origins and historical background of the River Severn and of the trows which were used in the intensive navigational activities of the River was described, together with the background and details of the sole known remaining trow *Spry* which illustrated just how important and significant this, remarkably almost intact, survivor from a great era was.

These factors eventually caused the USNT to feel that the restoration of *Spry* might be a suitable project to fulfil their main initial objective, as an alternative to constructing a replica. While *Spry* was an estuary trow and was too large to navigate in the upper River, it was basically the same in design and in fact did have connections relevant to the USNT's sphere of interest in the upper River, in that some such larger types of trow were actually built in Ironbridge. These were floated down to the lower reaches and fitted out there, never to return. In 1981, therefore, the USNT Management decided to investigate the possibility of the Trust becoming involved with the *Spry*.

Initially informal contacts were made between Mr. H. John Brown, representing the STPT, Kenneth Severn of the London (fund raising)

Committee of the IGMDT, who had many useful contacts, and Mr. Stuart Smith, then Deputy Director of the IGMT and joint Secretary of the USNT. It was agreed that the USNT would consider taking over the recovery and restoration of the vessel, subject to an expert examination of its condition.

Accordingly the STPT arranged for *Spry* to be dry docked in the dock across the Diglis Basin from where she rested on the mud, and to be examined by Mr. R. S. Hatchard of British Waterways, the Dock-keeper at Diglis. After applying plastic sheeting patches over the worst obvious holes around the hull just below the water line, *Spry* was pumped out using a 6-inch pump, a task which took one and a half hours. The examination took place and was reported on the 8th November 1981. Because the hull was not cleaned out of debris and mud, it was mainly an external examination. Mr. Hatchard's report confirmed that the deterioration was quite advanced. He estimated that 90% of the planking, much of which had been protected by thin tinned sheet cladding below the water line and elm sheathing around the unladen water line, would require renewing. Virtually only the elm keel and the pine keelson were found to be on good order, possibly because they were not the originals. These were, however, key members holding the hull together. His general, and perhaps rather obvious conclusion, was 'I am afraid to say that the *Spry* is just a rotting hulk, quickly returning to nature'. (The report in full is included in the Appendix.)

While rather pessimistic, the report did not indicate any drastically worse situation than expected and therefore it was decided to recommend acceptance of the vessel and the investigation of the costs and task of restoration, to the USNT Management Board, at a meeting already convened for the 16th November 1981.

The minutes of the meeting, which record a detailed discussion of the situation background and possible actions, are felt to be sufficiently interesting to justify inclusion here in full:

Minutes of Meeting of the Upper Severn Navigation Trust Management Team

Held at the Wharfage, Ironbridge, 7.00 p.m. Monday 16th November 1981

Present

1. *Lt. Col. Sir John Corbet (Chairman)*
Mr. Richard Barker
Dr. Neil Cossons
Mr. Michael Darby
Mr. Selwyn Devey
Lt. Col. A. P. Sykes

Mr. Peter Turnbull
Mr. Brian Waterson
Mr. Stuart Smith)
Mr. Mark Harwood-Little) Joint Secretaries

Apologies

2. *Mr. E. Bruce Ball*
Mr. M. E. Braine
Mr. N. Clarke
Col. A. P. de T. Daniell
Dr. B. Greenhill
Mr. D. Lawrance
Dr. B. Trinder

General

3. *The meeting had been called to review the situation on the Spry, outlined at the meeting on 13th July. Options were still being studied, but as the Chairman was going abroad for four months, it was important to review the situation and agree further action.*

Previous Minutes

4. *The previous minutes were not discussed, as they recently formed a background to current action, which was to be reviewed.*

Discussions with the Severn Trow Preservation Trust

5. *The Severn Trow Preservation Trust (STPT) is a Charitable Trust, formed to preserve and restore the Spry, believed to be the last Severn Trow. Considerable alterations had been made to the hull of their Chepstow built 1878 (actually 1894 – Author) craft, in its lifetime, 71ft × 18ft beam, designed for use in the lower Severn. She lies sunk in Diglis Basin, Worcester. The STPT were finding progress difficult, and had concluded that it would be better for a Trust located on the River to take over the ownership and responsibility of restoration. They held a figure of between £1,200 and £1,500, free of debt. They had tried to find a suitable group in the lower Severn to take over the Spry, as their aim had been to restore the craft and sail her in suitable water. Having failed in this, they had approached Mr. Stuart Smith, unofficially. He had explained that the USNT, in conjunction with the Ironbridge Gorge Museum Trust, hoped to moor a Trow – original or replica – on the Severn. Mr. Smith had agreed that if the Spry were taken on by the USNT, the aim would be to use her in that manner, keeping her name and design. The possibility of amalgamating the Trusts could be persued, but in any case the involvement of STPT members would be welcomed. They were happy to transfer this sum if a suitable agreement was reached. The W. A. Cadbury Trust had subscribed much of the money, and they would need to be satisfied, as would the Charity Commissioners. Mr. E. Thomas was persuing the legal position, without*

commitment. It is not clear whether Spry is currently owned by Mr. Brown, or by the STPT.

Suitability of the Spry for USNT plans

6. *The craft used in the upper Severn in the 18th and 19th Centuries included Trows (a word often used loosely, but if used with precision, meaning a craft capable of navigating the lower tidal portions of the river) of 80 tons burthen (i.e. freight capacity), with a similar plan view, but shallower draft than the Spry. (The latter might draw 5ft to 6ft laden, upper Severn craft drawing maybe 3ft 6 inches).*

Craft similar to the Spry were built in the Gorge, but plied in the lower reaches, returning occasionally to refit. The William built in the Gorge in 1809, which sank in the Bristol Channel in 1939 is a well documented example.

7. *On these grounds it was thought that it would not be historically misleading to display the Spry in the Gorge.*

A replica was inevitably second best, but if the Trust were to build a replica, she would have been shallower in draft, and lack of knowledge of certain details would make arithmetic replication impossible, and, perhaps, smaller for reasons of economy and handling.

Possible Actions

8. *Discussions of possible actions and their costs were necessarily limited because of lack of sufficient information.*

9. *Cash and resources available, or potentially available, were:*

a. The Manpower Services Schemes resources and very limited funds.

b. A possible further £2,500 to £3,000 from the STPT and the Science Museum, subject to the USNT satisfying both, and the W. A. Cadbury Trust, as to their plans. (The Science Museum could grant aid at 50%, up to £20,000).

10. *The Spry had been floated into dry dock at Diglis, (lifted in 1½ hours with a 6-inch pump), and a survey had been carried out by Mr. Hatchard on 8th November 1981, a copy of which is attached, (with photocopies of photographs). Some 12 to 16 patches were applied, mostly to the port side, and the Spry was again allowed to sink. Time was limited to 2 days in all, as the dry dock was wanted for another craft.*

The survey suggests that the craft is best left submerged at this stage to avoid further deterioration, and that much of her planking will require replacement. The planking will however provide templates for replacement timber and ensure that the planking is correct.

11. *The state of the fastenings is not known – though Mr. Richard Barker's comments on hulks on the river suggests that the iron may well be sound, the wood rotting around it. The ceiling must be removed to ascertain this important fact, but this would need the agreement of the owner.*

12. *The lines of the craft have not been taken, (nor was the waterline*

noted accurately when the Spry was docked. Her draft was estimated as 3ft to 3ft 6 inches, but this would need to be measured in several places to allow for her hogging when floated unladen). There is thus currently neither information for the record, making a replica or model possible, nor is her current weight known beyond an intelligent guess. Displacement plus lines are required, plus timber drawings and sizes as a check on calculation. Mr. Barker said that a difference of 5 tons could make a considerable difference in cranage costs (and will make a difference in draft of between 2 and 2½ inches – hence the need for accurate measurements).

13. *To tow the Spry upstream, the 36 miles odd to Ironbridge would require detailed reconnaissance of the river to determine depths and flow rates and shallow stretches when in spate. The view taken was that the job might take months, requiring considerable resources to be available at short notice when the water level was high, the Spry being moored and sunk at intervening periods. The Army would be most unlikely to be interested in such a prolonged and uncertain timetable. Comparatively little was know of the river, and there would be security and insurance problems. On the other hand, such a project could bring publicity and sponsorship, and would eliminate the need for expensive cranage, but would lead to the problems of preparing a riverside site and security mentioned below. The meeting was inclined to discount this option, on the information available.*

14. *This leaves the use of road transport for part or all the journey. The possibility of lifting the Spry out of the water at Diglis Basin was discussed, though might well be impracticable. (The weight being estimated at 80 to 100 tons, but see para. 9), as well as towing the craft up or downstream to a suitable point, lifting her out, and taking her by road.*

15. *Suggestions that the Spry could be hauled up a slipway onto a trailer, were regarded as impracticable.*

16. *In view of the statement in para. 7 that much of the planking might need replacement, it was suggested that the craft might either be cut in two (is the 16-inch keelson scarfed ?), or stripped right down and transported as marked timber. Or if it was worth transporting the whole hull, would it be sensible to remove the ceiling – and perhaps decking – to save crane weight?.*

17. *Dr. Cossons said that if the Spry came to the Gorge by road – whether whole or in pieces – it was probably best to carry out the restoration in the Blists Hill Open Air Museum. Security and workshop resources would be provided most economically there (though the latter would need improving). The snag would come later, in perhaps three years time when the Spry might have been restored. She would need taking down to the river bank and launching, while mooring facilities at the Severn Warehouse site, together with a nearby slipway, would be required. A proposal that the Spry could be jacked up and a trailer built for her economically*

to avoid the need for cranes, was thought to be impracticable in today's world.

Costs

18. *Guessed estimates to date, reported by Mr. Brian Waterson are:*
    a. Road Transport (Wrekin Road Transport) excluding cranage
    (1) With a double bolster ('trombone') £600
    (2) Flat bed trailer £2,600

It was thought that the Spry would not stand the cheaper 'trombone' system in her present condition.

She was in any case, too high, and some 18 inches would have to be removed from the top of the Transom and Prow, for bridge clearance.

b. Cranage. This would depend on site and weight. Discussions so far had been with Sparrows Ltd., whose yard at Gloucester closes at Christmas, and is sold for a project which will block access with buildings in due course.

The Sparrows site at Gloucester is not suitable for the use of two small cranes, and would require the greater cost of a large crane at more than minimum reach.

19. *Allowing for river transport, with a neoprene skin, tyres and pumps, and lift on and off transport at each end, total transport costs this way were estimated at between £12, 500 and £20,000 on such information as was available.*

20. *Mr. Thomas has spoken to Sir Frank Price, Chairman of the British Waterways Board, who had docks on the Severn, including Gloucester and Sharpness.*

21. *Total costs would include restoration, which might only involve £20,000's worth of timber plus the MSC labour, re-launching, mooring, slipway, and revenue costs.*

NOTE *Subsequent discussion suggested that, if static cranage is possible at Sharpness, it may be possible to use air bags and/or jacking at Blists Hill and avoid the use of cranes altogether. Possible company support on this was discussed between Sir Peter Gadsden, Mr. R. A. Joseph, Dr. Neil Cossons and Mark Harwood-Little on 5th December.*

Policy

22. *At this point the Chairman said that before the USNT could consider taking on any responsibility for the Spry, and by implications, for seeking a considerable sum of money, it was important to get opinions as to her value both to the Nation and to Ironbridge. Was the craft of sufficient importance for the transport costs to be justified, assuming the money could be raised? The USNT had been created as an extension of the Ironbridge Gorge Museum Trusts, and their reputation would be affected by the actions of the USNT. Dr. Cossons agreed to:*

a. Discuss the status of the Spry project with Dr. Greenhill,

Director of the National Maritime Museum, whom he would be seeing on the 8th December. If it was appropriate, the case could be put to the Maritime Trust, who could consider it against similar projects elsewhere, and could give it an objective priority as well as either providing or encouraging financial support. Dr. Greenhill may also be able to help with further technical and costing advice (as has already been arranged on the replica project, with Mr. Alan Vinear's visit).

b. Discuss the project with Mr. Trevor-Jones, as Chairman of the Museum Trust, in relation to its priority at Ironbridge. At the moment, the part played by river transport was only shown to a very limited extent in Severn Warehouse. The Severn was reputed to be the second busiest river in Europe in the seventeenth century (to the Meuse), and its significance until the arrival of rail was enormous.

23. *The Chairman was anxious to see that the revenue aspects of the project were measured and taken into account. Would it be better to build a scale model at between ¼ and ⅛th scale, say? Or should the Spry be exhibited dry? It was thought that the STPT might object to the latter, and Mr. Barker argued that the craft would need covering if this was the case.*

Further Action

24. *If the project might continue, it would be necessary to put her into dry dock again for further investigation, to provide information outlined in paras. 9 and 10, which requires the necessary technical expertise, and to answer any questions raised by Dr. Greenhill. STPT agreement would be necessary for some action, and it could be appropriate to propose to carry out certain work at their cost, as their agents. Full discussions with the Severn Trent Water Board and with the British Waterways Board would also be necessary.*

Fundraising

25. *Fundraising was not discussed except as above, the costs and rate of expenditure not having been established.*

Other Business

26. *The Treasurer, Mr. S. R. Devey, was given authority to pay the Auditors Bill of £40.25 from the bank account of £150.*
27. *Mr. Darby suggested that some action be taken to get an appropriate small craft in the river, both for use and to start telling the story. Could this be done with MSC help?*
28. *Subsequent meetings would be called as required by Dr, Neil Cossons, in the absence of the Chairman until April 1982.*

Signed — *Lt. Col. Sir John Corbet,*
Chairman.

The Report of Spry's condition, by Mr. R. S. Hatchard is included in the Appendices.

The actions agreed in the meeting were pursued and it was confirmed after discussions with the Directors of the National Maritime Museum and the Chairman of the Ironbridge Gorge Museum Trust, that the vessel's preservation was of sufficient interest Nationally and to Ironbridge to justify a major effort in fund raising and provision of facilities to achieve the objective.

Under the enthusiastic direction of Dr. Neil Cossons, the Director of the IGMT at that time, actions were accordingly put in hand to initiate the project. One of the first priorities was for the USNT to acquire the legal ownership of the vessel from its current ownership vested in Mr. John Brown of the Diglis Boat Company. This was a fundamental condition stipulated by the Maritime Trust, before they could consider a grant of funds, which they indicated they were likely to be able to contribute. The transaction was concluded and a Deed of Transfer signed on the 10th December 1981, at the very nominal figure of £1.00!

The means and costs of removing *Spry* from Diglis Basin to the major Ironbridge Gorge Museum site at Blists Hill were further investigated.

Floating and hauling the craft upstream, even with a supporting raft to reduce the draft, was confirmed as being completely impractical. The possibility of dismantling her partially or completely for transportation was also considered and rejected. After discussion with specialist heavy lifting and transporting Companies, the remaining option of attempting to lift the vessel bodily out of the water for road transport was eventually decided upon. Here too the options of floating *Spry* down river to Gloucester or Chepstow, to utilise wharfside fixed cranage and hopefully minimise cranage costs, were considered and also rejected.

After further discussions with the lifting and transportation specialists, it was decided that a large enough mobile crane could be erected alongside the inner basin at Diglis, where there is an adjacent suitable road. Using a specially designed lifting frame to support the ancient and rotting timbers of the hull, *Spry* should be capable of being lifted from a floating position alongside the wharf, directly onto a low loader trailer. Further, using a trailer with built in hydraulic jacks, and supporting her on loose steel cross beams on the trailer, it would be possible to avoid the expensive repeated use of a crane at the destination, Blists Hill, by using the trailer jacks to lift and lower *Spry* onto ground supports. The low loader trailer also could be of a type with a well between two groups of axles front and rear, which would just keep the overall height of the load on the road low enough at 16 feet, to pass under bridges across the M5 Motorway, which was the direct route to the Telford area.

The cost of this exercise was estimated by the Museum Engineer,

Brian Waterson, on the more recent information now available, to be approximately £13,000.

With this basic plan agreed arrangements for finalising details and for raising the necessary initial funding, were put in hand in early 1982.

Clearly, the very first task, the lifting of *Spry*'s ancient and delapidated timbers, in one piece, from their resting place on the mud of Diglis Basin was going to require very careful handling and expert assistance.

As a result of a recommendation made by Kenneth Severn of the IGMDT, in January 1982, the Civil Engineering Consultants Archibald Shaw and Partners, based in Chichester, Hampshire, were approached for their expert advice on the best means of supporting *Spry* during the heavy but delicate lifting operation. Mr. Archibald Shaw, a former Director and Chief Engineer of Tarmac Construction, had a wide experience in a variety of Civil Engineering projects before and since starting his own business after retirement from Tarmac. He was also a keen yachtsman, being the owner of a Swedish built 41-foot ketch and was readily interested in the project. While initially sceptical of the proposal to transfer *Spry* by road to Ironbridge, either whole or in parts in view of the report on her condition, he willingly agreed to make an assessment, in conjunction with a personal friend, John Bartlett, who was an eminent nautical expert.

The initial step in this was to move the vessel into an adjacent dry dock where she could be examined in detail, to form a decision on the best way to lift *Spry* out of the basin onto the low loader for transport to Ironbridge. This was arranged for the 15th February 1982, with a limit on the dry dock usage time due to a heavy demand for it at that time of year.

The worst gaps in the planking were again sealed by manoeuvring plastic sheets through the mud, the water pumped out and the craft poled across the basin to the dry dock where she was exposed for the detailed examination. This time the mud and debris were to be cleaned out. The initial efforts by a Manpower Services team did not make fast enough progress in view of the dock availability time limit but the Waterways Board gave very effective help to speed completion.

Mr. Shaw accompanied by Mr. Bartlett, visited the Diglis dry dock on the 19th February. They found the craft to be in rather better condition than they had expected from reading Mr. Hatchard's earlier report, which had necessarily been confined mainly to the exterior. The removal of the mud and debris enabled the keelson and its fastenings to be better examined and they were impressed and encouraged by the confirmation that this was intact and in remarkably good condition. This massive timber ran the length of the bottom of the hull, acting as a spine to the vessel, sandwiching the bottoms of the side frames to the external centre keel, which while scarfed (joined) in the centre, was also in good condition. By means of a thorough probing with a sharp knife, the keelson was found to be 'as hard as iron' and showed hardly any deterioration.

Notwithstanding expressing the opinion, after this examination, that they felt it would be folly to renovate the craft and that it would be better to preserve her in situ, or to produce a replica, Mr. Archibald Shaw, in particular, was now convinced that the keelson would be a strong enough support to enable the vessel to be bodily hoisted out of the dock onto a low loader, using a specially designed lifting frame, in order to persue the policy already agreed to transfer *Spry* to Blists Hill. He further undertook to design a suitable frame and, most generously, as a free donation of his expertise and time.

Even with the substantial keelson, it was still essential to distribute the lifting loads evenly over a sufficient length of the hull, which could be done by using a double articulated suspended frame. Such a lifting frame was also included in the quotation from Sparrows Crane Hire, dated 11th March 1982, which followed their visit to examine *Spry* while dry docked and to discuss in outline the method of lifting. The articulated suspension principle suggested by Mr. Shaw, appears to have been mentioned to them during the meeting.

Mr. Shaw's recommendation of the use of a frame based on this principle was based on his experience of his use of the same articulated design in reverse for his invention of the 'Shavian Finisher' which was a former breakthrough in concrete Motorway surfacing technology – a remarkable coincidence of modern transport technology being allied with its ancestral origin in river navigation.

The design for *Spry*'s lifting frame provided a rectangular member to be suspended horizontally from a single crane hook by four 20-tonne wire slings attached to each corner. A further four straight lifting beams suspended freely by a cable at each end, from each of the four corners of the frame, giving in total 8 freely floating suspension shackles to attach to straps under the hull.

The 23-foot square main frame was constructed from 12 × 4 inches rolled-steel channels, bolted together and cross braced. The frame width was designed 5 feet wider than the hull to minimise the inward crushing load during the lift around the open top of the hull. The frame length and the suspended beam lengths, which were of H section rolled steel, 10 × 6 inches in section and 15 feet long, gave four suspension points approximately 10 feet apart along each side. The lifting load was therefore distributed along 32 feet of the hull amidships to within 18 feet of the bow and stern. Mr. Shaw calculated that the keelson was strong enough to carry the cantilevered loads at either end.

The drawings for the frame from A. Shaw and Partners were sent to the Museum on the 17th November 1982 and its fabrication was carried out in the Museum Trust workshops. Due to the cheap availability of sufficient steel channel of 15 × 4 inches section, this was used by agreement, instead of the designed 12 × 4 inches section for the main frame. The specialist items, shackles and wire slings etc. were purchased from Holt Williams and Co. Ltd., of Powys. They also supplied the four 12 inch wide by 11 metre long Paraweb woven nylon slings for cradling the vessel in suspension from the lifting frame. The material costs for making the frame 'in house' to the donated Shaw design were £2, 840, plus £300 for proof load testing. This compared with £6,000 quoted from Sparrows for their supply of a similarly designed frame, a significant saving.

In the meantime from the updated quotations received, Sparrows Crane Hire were chosen to supply the heavy duty (200-tonne) mobile crane to lift *Spry* out of the water onto a low loader to be provided by the local firm, Wrekin Transport Services Ltd. of Bridgnorth. The cost of the crane hire was £2, 500 plus VAT and the cost of the trailer and delivery to Blists Hill was £2, 200 plus VAT. The total cost of lifting and transporting *Spry* to Blists Hill was therefore £7840 with the very generous donation of the free of charge design of the lifting frame, by Archibald Shaw and Partners.

This cost was adequately covered by the funds likely to be available, which were assessed in November 1982 to be at a potential total of £12,500, from the following sources:

	£
Heart of England Tourist Board	2,500
From *Spry* preservation Society	1,200
Ironbridge Development Trust Reserve	1,050
Promised by P. H. Holt Trust	1,000
(conditional on £3,000 being raised elsewhere)	
Promised by the Telford Hotel, Great Haye	500
Science Museum Grant	6,250
	12,500

All arrangements were therefore put in hand and the date of Saturday 26th March 1983 fixed for lifting *Spry* from her muddy incarceration.

It was arranged with British Waterways that the vessel, which was again almost completely submerged at her original berth, would be pumped out again and refloated on the Monday 14th March and be moved across into the dry dock for preparatory treatments. She would be floated out of the dry dock and from the main dock into the inner basin on Friday the 25th March to be lifted out the following morning.

In the meantime, a suitable route from Diglis Basin, through Worcester and up to Blists Hill, Ironbridge, was researched in detail by Brian Waterson which meant driving over the alternative routes and measuring all the many bridges for height clearance. This he did with a pole with the top of his car as reference. The following route was eventually decided upon, using the M5 to Junction 4 then via the Kidderminster and Wolverhampton by-passes and on the A41 and A5 to Telford and Madeley. The detailed route was as follows:

Leave Diglis Basin via Diglis Road
Join A38 Bath Road, (by Bristol Street Motors)
BL A38 to traffic lights
TL Sidbury to traffic Island by the Cathedral
BL Deansway
BL Bridge Street
TR North Quay and continue Croft Road and Castle Street
TL The Tything at traffic lights
Continue Barbourne and Ombersely Road to A449,
Worcester Northern Link Traffic Island
2nd Exit at island to Junction 6 M5
(CAUTION Travel in overtaking lane under Spellis Green Railway Bridge at reduced height of 16 feet 11 inches)
TL M5 to Junction 5 (Rashwood)
(CAUTION When travelling under Junction 5 Bridge on A38, travel at reduced height of 16 feet 11 inches)
Circumnavigate Junction – A38 via Bromsgrove Eastern by-pass to M5 Junction 4
TL A491 to Hagley
TL A456 via Blakedown to Kidderminster
TR A449 to Wolverhampton
TL Wolverhampton Ringroad
Continue A41 to Pichmere Island
TL A5 to Priorslee Island
TL Stafford Park Road
TR to Randlay Island
TL Queensway (use slip road at Stirchley Island)
Continue Queensway
TL Brockton Way to Brockton Island
TR to Sutton Hill Island
TR Bridgnorth Road to Cuckoo Oak Island
TL A 4169
TL to Museum.

In due course on the 14th March, when it came to pumping out *Spry* again, it was found to be a much more difficult task than before, due to the rotted planks leaking at a more rapid rate following the two previous disturbances. Extensive patches of plastic sheeting were applied over the worst rotten areas on the outside of the hull, and fixed by wooden laths and nails. Nevertheless the pumps supplied by the Museum

Figure 3: Internal stiffeners fitted to avoid crushing during the lift.

still proved unequal to the task and assistance had to be sought from the local fire brigade. This was willingly given.

As something of an afterthought, it was decided that the bulwarks needed bracing. The lifting frame had been designed 5 feet wider than the top of the hull at either side to avoid an inward crushing force, but this meant that the slings would leave contact with the hull some 4 feet below the bulwarks, leaving the open top of the hull unsupported. Arrangements were hurriedly made to use cut down telegraph poles fitted into steel brackets clamped over the bulwarks to give support against both inward crushing or any tendency to bulge outwards. The brackets were urgently fabricated in the workshop at Blists Hill. The ends of the wooden poles were drilled across their diameters to take ¾-inch bolts to secure them in the brackets. (Figure 3) The wisdom of providing the bracing was evident afterwards as the steel bolts were found to be bent.

Special mention must be made of the considerable personal efforts of the Engineer-in-Charge, Brian Waterson, in all the preparatory arrangements and discussions with Surveyors, British Waterways and many others and in the planning, supervising of the activities and actually assisting in their physical execution.

In the midst of all the activities of the final preparations, on the Thursday the 24th March, consternation was caused by the appearance of two Senior ranking Police Officers, who asked to see the man in charge, which of course was Brian Waterson. The Chief Inspector and accompanying Inspector asked him what was proposed and after being given the details and viewing the site for the lift onto the low loader, pronounced that the loading clearly could not be achieved without closing the Diglis Road, on which the transporter would have to be parked for some time. This would be illegal without an Official Road Closure Authorisation and furthermore, any contravention could lead

Figure 4: *Spry* being refloated in Diglis Dry Dock, Worcester, after patching hull to limit water ingress.

Figure 5: Hand towing *Spry* into position for lifting her out of the water.

to the arrest of the Engineer in charge on a charge of obstruction! Brian was not aware of an application having been made and accompanied the Officers to the Police Station for enquiries to be made to the Worcester Highways Department. It was ascertained that no application had been made and the Authority spokesman was adamant that 14 days notice was required before one could be issued! After all the effort and at this late stage, this news was rather disconcerting to say the least!

To try to resolve the unexpected and difficult situation, the Chief Superintendent of Police was involved and he was very helpful in negotiating with the Highways Authority over the telephone, on behalf of the salvaging team, pointing out the embarrassment to all concerned with the very extensive Media coverage arranged, if a postponement were necessary with the lifting only two days away. Eventually a

Figure 6: The cantilevered lifting cradle designed by Mr. Archibald Shaw.

compromise was reached and permission given for the road to be closed at 6.00 a.m. Saturday morning, to cause the least disruption to traffic, and for *Spry* to be clear of the City boundary by 10.00 a.m. As this meant bringing forward the planned time by 5 hours, with the work already somewhat behind schedule, an extraordinary effort was called for.

Spry was duly refloated on the Friday and poled and pulled across the Dock into the inner Basin to the lifting berth. (Figures 4 and 5) Constant pumping was necessary to stem the inflow from numerous residual leaks. After working through the night to finalise the fixing of the internal bracing and preparing the lifting cradle, the preparations were completed just in time to meet the deadline of 6.00 a.m. Saturday morning, to commence the lift and the crane (Figure 6) and low loader were both ready. It had originally been envisaged that a diver might be necessary to make sure the slings were positioned correctly under the hull, but a decision to manage without one was justified, as no problem was encountered in holding the ends by one person on either side, looping them under the hull and moving them along into position.

It had been decided that steel channel cross girders would be placed on the low loader, with additional packing to level up a well between the two groups of axles. *Spry*'s bilge keels would rest on these during the journey and they could be jacked up at Blists Hill to allow the loader to be moved out from under the vessel, thus avoiding another costly crane lift.

The early hour and the unavoidably short notice of the change of timing meant that the media coverage was somewhat disappointing. Very shortly after 6.00 a.m., with varying degrees of trepidation in the participants and the onlookers who had managed to be there, depending on their confidence in the judgements of the strength of *Spry*'s aged timbers and the calculations of the stresses involved, the lift began. Slowly she rose further and further up in the water, drooping slightly, but perceptively more and more, at the bow and stern, (estimated at 6 to 9 inches) – and then she was clear – high but not quite dry, above the surface! Dripping but in one piece. (Figures 7 and 8) The keelson had fulfilled the confident expectations of the lifting expert Archibald Shaw.

Spry was weighed on the crane and was 48 tonnes, which, with the additional weight of the frame, was well within the 60-tonne lift

contingency allowed for. The wetness of the timbers contributed considerably to the weight.

Loading was from then on, merely a routine, with the lifting slings now doubling as securing straps to hold the vessel down. (Figures 9, 10 and 11) And so the convoy with Police escort in front and Brian Waterson in his car in the rear, with hazard warning lights flashing, were on the way. Brian's good warning intentions however, were soon stopped as he was warned he was committing an offence by using his hazard lights on the move – his second brush with the Law in three days!

The navigation of the low loader and its bulky passenger, through the narrow streets of Worcester and negotiating the numerous traffic lights, was a difficult task but accomplished without any disasters (Figure 12) and the convoy was clear of the City almost on time, just over-running the deadline of 10.00 a.m. by 20 minutes. From then on the journey was relatively plain 'sailing', for, by using the built in hydraulic jacks of the low loader and lowering the platform down to a minimum ground clear-

Right: Figure 9: *Spry* on the low loader.

Figure 10: Showing packing in the well of the low loader.

Figure 11: The start of the journey to Ironbridge.

OPPOSITE:
Top: Figure 7: *Spry* in the cradle, lifted clear of the water.
Bottom: Figure 8: End view of *Spry* suspended, showing droop at rear end.

Above: Figure 12: Negotiating a narrow street in Worcester, *Spry* almost dwarfing the buildings.

Left: Figure 13: On the M5 motorway.

ance on the Motorway, the vehicle and its load could pass under the bridges, which have a 16-foot clearance. (Figure13) At a very respectable 40 mph. the Frankley Service Station was soon reached where a temporary halt was called for a change over of the Police Escort, this being the limit for the Worcester County Constabulary and the start of that of Staffordshire. Entering Shropshire along the A5 road, North of Wolverhampton, the vessel and trailer negotiated another low bridge, Thomas Telfords aqueduct carrying the Shropshire Union Canal, with only inches to spare. (Figure 14)

Figure 14: Minimal clearance under the Telford-designed canal aqueduct over the A5, near Western-under-Lizard.

Figure 15: Evening arrival at Blists Hill.

Figure 16: Manoeuvring *Spry* on the low loader into Blists Hill Museum site.

Figure 17: *Spry* in position on the prepared hard standing.

From here on the journey was to expected to be reasonably straightforward, so at this point Brian Waterson went on ahead with the aim of having a wash and a short rest, having been up all night, but after a few minutes at home he had a phone call to say the convoy was entering the Telford area. Leaving immediately he found *Spry* was already in Legges Way, nearly at the Museums Blists Hill entrance, and was able to help, standing on top, in lifting or lopping the numerous overhanging tree branches and telephone wires in the overhung roadway past the main Museum entrance, down to the goods entrance behind the David and Samson blast furnace air pumping exhibit, where a hard standing for *Spry* had already been prepared. (Figure 15)

Manoeuvring the loader off the road was a very difficult task necessitating a temporary traffic diversion around the plantation between the two access ways onto the road, but it was eventually accomplished and *Spry* was located over the narrow concrete hard standing at approximately 5 p. m., just as darkness fell. (Figures 16 and 17) There, the next day, the loader platform was jacked up to maximum height on its in-built system and 2-foot by 12-inch timber baulks were placed under the cross beams carrying the vessel. The clearance was not, however quite sufficient to enable the loader to immediately be run out, due to the well in its centre. The vessel had to be independently jacked higher at one end to allow this, which caused some concern again regarding the strain on the

Figure 18: Celebrating a job well done. Brian Waterson and Brigadier Goulburn, second and third from from left front respectively, with the team of Museum helpers.

timbers and the risk of breaking its back, but all went well and once again the keelson proved an adequate support. By midday on the Sunday the task was finished with the transport departed and *Spry* firmly established, leaving all concerned feeling very satisfied with the completion of a very considerable undertaking and not a little relieved that everything had gone so well. (Figure 18)

CHAPTER 4

Obtaining the Timber

Quantities needed and sources of the timber

Brigadier C. Goulburn DSO, DL, a Museum Trustee, unfortunately since deceased, undertook the essential task of finding donors for the timber for the reconstruction, which at that time was roughly estimated at 500 cubic feet. Prominent in the landowning community in Shropshire, related to the Hon. James Hamilton, now Lord Hamilton of Dalzel, owner of Apley Estates and a friend of John Ransford, timber agent, the Brigadier had many invaluable contacts and in early 1983 he had a very encouraging response from initial approaches to other estate owners in the County. These indicated that most, if not all of the requirements as originally estimated from what proved to be an optimistic survey, might be obtainable from generous donations from these estates.

However it was not until 1986 that the USNT was properly in a position to commence preparations for the restoration and to quantify the amounts of timber needed, following the recruitment of Master Shipwright Alan Williams in the February of that year. Alan, an experienced young man, had followed his apprenticeship and initial practical experiences in wooden boat building and restoration in England, by gaining further valuable experience in this field as a journeyman, working in several European Countries, North and South America and finally for over four years in Australia. Alan was nearing the end of an assignment in Australia when he heard of the need for a Master Shipwright for the *Spry* project from enquiries through his parents in Shropshire. He made the approach to the USNT, to mutual good fortune, and was readily given the job to start on completion of his Australian assignment in 12 months time.

Alan's first tasks on taking up his appointment in Ironbridge, were to assess the magnitude of the work involved and then to draw up an outline programme and estimate of the resources needed including the calculation of the timber requirements.

The first reports on studies carried out while *Spry*'s timbers were still wet were found to be rather optimistic in the amount of the original boat which could be saved, bearing in mind the USNTs clear terms of reference to restore the boat to river-worthy condition. This was due to the fact that in the intervening years in storage at Blists Hill, Ironbridge, the timbers had dried out and the full extent of the deterioration was more clearly revealed. It was apparent that none of the oak frames were sound enough to be retained and that the few planks which seemed sound were in fact rotted around the bolts and nails which would prevent their reuse on new frames.

The task faced, therefore, to fully achieve the objectives, was no less than the complete rebuilding of the vessel with new timbers almost throughout, using the old planking as moulds to make spiles (templates) for the new frames and planks.

The volumes of the different timbers required were taken from the dimensions of the originals, the principal ones being the 40 double sawn oak frames and 2,000 linear feet of larch planking. This was doubled to allow for process wastage in sawing. Later the wastage allowance was increased by another third for the oak framing. The cubic footage of solid timber thus arrived at was converted into an approximate number of trees needed to provide the amounts, using the Hoppus Decimal Tables of solid content to round timber equivalent.

The system of measuring round timber to determine the volume of useable timber, after allowing for sap wood etc., is based on measuring the girth of the standing tree at chest height so as to clear the buttress roots near ground level and multiplying by the length of the estimated useable height. The tables in current use are named after Edward Hoppus who introduced them early in the eighteenth century as a decimalised improvement over tables already in use for a considerable time before that. This was long before decimals were in general use in Britain. While a measurement of volume, the result is expressed as 'Hoppus feet'. The result is a useful approximation for trees of normal taper and sound, but the yield can be affected by tree defects, some of which may not be visible on the exterior, such as 'brown rot' or 'shakes', (internal splits). These can be star shaped from the centre or radial around the growth rings from shock or damage during growth or during felling from a hard fall, when they are called 'thunder shakes'. Even with the great care taken in subsequent felling there is still loss due to 'thunder shakes'. However the estimate had to be based on something and the Hoppus Tables gave a reasonable basis.

The choice of timbers around the time of the original construction were the traditional English oak (Quercus Robur) for its strength in the frames, pitch pine for the planking above the water line, for its flexibility, dense fine grain and relative freedom from knots, but requires care to avoid splitting, and elm for the keels and planking in permanently submerged applications, for its good lasting properties under water and due to its ability to resist knocks and abrasion.

The situation at the time of *Spry*'s rebuilding had changed somewhat, 90 years on. Pitch pine was difficult to obtain and expensive in the quality and quantity required, from reasonably local areas in Shropshire and Staffordshire. Elm, due to the ravages of Dutch Elm Disease, was virtually unobtainable. For the planking in the reconstruction therefore larch was opted for, which was available in very good quality in Shropshire. Thus the requests for tree donations from the local estates were to be primarily for oak and larch to cover an estimated requirement of about 2000 cubic feet of shaped oak and 2000 cubic feet of larch, which from the Hobbus Tables equated to 20 oaks and 40 larch trees. In the event the indication of the Hobbus Tables underestimated these requirements as the Tables are based on normal commercial yields, which did not apply in the case of the *Spry* timber, especially the highly shaped oak planks. Therefore ultimately 28 oaks, 41 larch and also 13 Douglas fir trees were needed – plus one Cedar of Lebanon, as mentioned below.

With this information, Brigadier Goulburn was able to circulate 44 local landowners in Shropshire, adjacent areas in Staffordshire and Powis, in a letter dated 8th May 1986 (copy included in Appendix 4). The original generous interest and promises were fulfilled in good measure with over half of the landowners being able to offer what appeared to be suitable trees to donate, adequate in total to cover the revised estimated requirements.

One of the early offers was from the Attingham Park Estate, near Shrewsbury, which is owned by the National Trust. The tree offered was in fact a very fine Cedar of Lebanon, measured to have 200 Hoppus feet of useable timber, useful for its great length and durability for the vessel's hatch coaming, where particularly hard wear was experienced in the original duty of *Spry*. Being in the National Trust, itself a charitable fundraising organisation, Attingham were not able to donate the tree free but made a very nominal charge. The very first tree to be acquired for the project, therefore, was for one of the last requirements on completing the hull and was an actual purchase for a nominal £100. Two other purchases were also subsequently made totalling four oaks, but apart from these all the timber was donated to the Museum.

A list of the names and locations of these generous donors of trees, including those purchased from other Trusts at nominal figures, is included in Appendix 4.

The timber extraction

It was important that all the trees and locations were examined to make the final choices of suitability and accessibility. Alan Williams therefore visited all the various estates, making the selections with his knowledge of his requirements, particularly for the oaks to be fashioned into the frames, many of which needed to be very extreme in shape with 90° bends, these being the most difficult to procure.

Assessing the locations was no less important than assessing the suitability of the trees in relation to the accessibility for the heavy equipment likely to be involved. This was from the point of view of the topographical features – adjacency of tracks and roads – few of the trees were near a road! The softness and slope of the ground and the interference of streams, ditches, railway bridges and power cables, all had to be taken into consideration. Another very important aspect where trees were situated in areas cultivated for crops, was the timing for access. Some of the trees were in grazing land which could be accessed whenever reasonable weather conditions prevailed. Others, however, were in arable land with crops of wheat and barley, Winter and Spring, oil seed rape and grass for hay. In these situations access would have to be planned in the interval between harvesting, re-ploughing and the sowing of the next crop. About half the trees, some estimated at 40 tonnes standing weight, required a large, 10-tonne mobile crane and this also had to be taken into consideration.

This detailed survey took, on and off, three months to complete and when all the complexities had been assessed and the final choices made, it was established that there were 20 locations involved. These were marked on a large scale map of the area, together with the status of each, which formed a field plan and the basis for action. The locations, as can be seen from the map reproduced in Figure 19, covered a wide area of Shropshire and included the adjoining part of Staffordshire.

Ten of the landowners involved were able to undertake the felling with their own facilities, also the clearing of the cord wood and the transportation of the logs to the nearest road for collection, which was a very considerable additional help. In the other cases where the landowners did not have the facilities to offer, arrangements had to be made with specialist contractors as it was clearly not a task within the capabilities of the facilities of the IGMT which the USNT would otherwise have sought to employ.

This situation was assessed by Alan Williams in his note of 2nd May 1986:

1. The Museum (USNT) has little money to spend on outside transport or Capital Equipment.
2. The Museum lorry is very limited to 20-foot lengths, 3 tonnes maximum load, with no lifting gear and an inefficient winch.
3. The trees we have been given are usually in difficult positions: a) poor access; b) on hillsides; c) on soft ground.

   All this is is in an area not set up for or used to, boat and ship demands.

Following early favourable responses to Brigadier Goulburn's letter, Alan's attention turned to the problem of the eventual conversion of the timber into the required sizes for the boat building operation. The large size and awkward shapes of the oaks in particular, would present

Figure 19: Map of Shrewsbury area, marked with locations of majority of sites of donated trees.

a problem to sawmills even with large enough equipment, who were used to only producing straight planks.

Here the Brigadier's involvement with the timber industry provided an invaluable contact. Through his donation of oak towards the needs for the repair of York Minster, he had met Mr. Rodger Venables whose sawmill in Stafford specialised in sawing oak. The mill also had adaptable

Figure 20: Typical oak tree before felling.

equipment capable of handling logs up to some 5 feet in diameter and up to 45 feet in length. This seemed adequate at the time. Mr. Venables undertook to carry out the sawing of the oaks and as a gesture, also to saw the larch, quoting a favourable price of £28.50 per hour on a semi automatic bandsaw. Contact was also eventually made with the Kingsland Saw Mills Ltd., South of Ludlow, who agreed a similar arrangement.

Because the tree locations were in ones and twos scattered many miles apart, it was decided that those felled adjacent to the River Severn above Bridgnorth or North and East of the river into the Staffordshire border, covering 11 sites in all, would be transported to the Venables Mill in Stafford. Those in the South of the area, the remaining 10 sites, including one near Welshpool, would be transported to the Kingsland Saw Mills.

With the trees selected and facilities secured for the sawing of them after felling, arranging for the actual felling and transportation of those to be the responsibility of the Museum, could be put in hand, using the location and access phasing plan. Here Alan Williams had the good fortune to make what later proved to be a most valuable contact at the Three Counties Show in June 1986 at Shrewsbury. The Volvo stand had a heavy duty tractor and trailer for timber hauling on exhibition. The machine had been fitted out for Benbow Brothers Ltd., a round timber firm. Enquiries and discussions led to a satisfactory contract with Benbow Bros. In July 1986 the first tree was felled at Morville near Bridgnorth on the Estate of Mr. Christopher Thompson. The operation continued as weather and crop harvesting permitted and most of the felling was completed by October 1986.

Very many difficulties were encountered and surmounted by the skill and ingenuity of Benbows staff, the tree feller and crane driver, Mike – who had a penchant for 'magic mushrooms' and sought them in all his spare moments – who could almost perform miracles with the crane in achieving difficult lifts and Alan, the tractor driver, whose skill in manoeuvering the bulky articulated vehicle was truly remarkable.

Mike's skill in felling the trees (Figure 20), which were often unbalanced by lop sided branches and situated on difficult slopes, was such that there were only one or two occasions where the useful part of the tree was shattered in the fall. Some of the trees were so big in girth that the large chain saw would not reach through. Where possible the trees on slopes were felled to fall uphill to lessen the impact or cushioned in falling against adjacent trees. In most of these cases where the felling was done by the USNTs contractor, because there were no Estate facilities, the landowners did arrange for their staff to clear up the debris. As this could amount to half the weight of the tree and take up to two days, this was very welcome assistance.

The tree locations, even of these selected as being reasonable of access, were nevertheless sometimes very difficult to reach with such a large articulated vehicle, along narrow twisting lanes and tracks, but due to Allan the driver's skill, he never failed to reach the objective somehow. Only on one occasion, in negotiating a newly laid stone hard core covered track, the tractor bogged down off the edge as it was turning. Undeterred and using a bit of help from the crane in lifting the trailer and pulling the tractor back onto the track, what seemed to be a catastrophe was overcome. On grassy slopes tractor wheel spin was often a problem. One of the first logs to be recovered, the Cedar of Lebanon from Cronkhill House, called for a high degree of tractor manoeuvering skill. The tractor

Figure 21: Oak log being loaded onto lorry trailer.

Figure 22: 'Dismembered elephants' ready for transporting to the saw mill.

and trailer had to be backed through two ornamental iron gates only fractions of an inch wider than the vehicle, and over a cattle grid of dubious strength, in negotiating a drive almost half a mile long. This was accomplished at some speed without detriment to vehicle or the gates!

The skill in operating the crane was no less impressive. This was particularly demonstrated at Morville. Two fine large trees were felled, in one of which Alan Williams counted 165 rings before he lost the count in the sap wood! This tree was well over the 10-tonne capacity of the crane. The only way to move the log to the nearest point the tractor could reach was to reel in the crane cable attached to the log until the cranes front wheels were lifted 6 feet off the ground. The hoist brake was then released momentarily allowing the front axle to fall and then applying it again hard as the axle was still falling. The resulting snatch jerked the log forward. This was repeated many times in a kind of 'bunny-hopping' operation until it was brought to the tractor. It says a lot for the crane and winch makers that no harm to the equipment resulted! With some of the trees like this one, it took up to a couple of hours to manoeuvre them onto the truck, one end at a time, due to their heavy weight and awkward shapes. Figures 21 and 22 show logs being loaded onto the trailer and a completed load ready for transporting to the sawmill.

There were many amusing and some hair raising experiences too. On one occasion the direct route to recover logs from an Estate at Ludlow was along an Estate road and over an old river bridge with a three-ton weight limit. The only alternative was a 20 mile detour through Ludlow town centre. It happened to be market day and manoeuvering the articulated vehicle through the crowds, who were in no hurry to make way and past the medieval timbered buildings with inches to spare was something of a nightmare.

The sight of the oak logs when they eventually arrived at the Sawmills caused great amusement to the staff there, used to straight tree trunks and they were likened to 'dismembered elephants'!

To ensure that the unusual requirements were fully met in the tree felling, Alan Williams was always present and also gave assistance as one of the team. The whole exercise of tree felling and recovery of the timber was no less difficult than foreseen but was very successfully accomplished with many such exciting moments as the few described. As Alan said afterwards, 'this was really getting down to basics in boatbuilding in a way I have never done before.' It was certainly a far cry from usual Museum activities!

The sawing of the timber

'Dismembered elephants' and the more normal looking logs started to arrive at the sawmills in August 1986 and this continued until after the end of September. The timber was stored outside in the sawmill stockyard until it had all accumulated and made an impressive sight. The quotation had been made on the basis of a continuous allocation of machine capacity and it was necessary to have all the timber available because of this. Sawing commenced in October 1986.

Each log required to be assessed first in relation to the optimum number of 2-inch thick planks which could be cut from its diameter, disregarding the rim of sapwood. The minimum number was three.

The handling and setting up of the logs with their very awkward shapes and the need to ensure the best alignment to produce the optimum shapes and lengths, was difficult and time consuming. In many cases awkward decisions were needed in chain sawing off excessively protruding stumps from the logs in order that the logs could pass through the 5-foot width of the machine frame, without sacrificing too much length in the resulting shaped plank. The longitudinal – fore and aft in nautical parlance – adjustment was fairly straight forward, but the radial adjustment was more critical as was the clamping of the logs in the correct position. This was accomplished by winding in the machine table clamping 'dogs' by hand with the log wedged in the position required. To ensure that the required shapes were produced, Alan Williams was again always present, as in the felling.

The yield of sawn plank weight from the stump weight was very low by normal commercial standards, and was estimated to be only of the order of 50%. This was due to the requirement only being for 2-inch thick planks, whereas commercially a variety of plank thicknesses would have enabled a greater yield to be achieved.

The heavy duty bandsaws were 9 inches wide, reducing with repeated sharpening, and 42 feet in circumference. In spite of high pressure washing and close scrutiny of the logs prior to sawing, occasionally the blade was damaged by lumps of metal or stones embedded in the wood, the tree having grown round them at some stage.

On a good day 8 to 10 logs were sawn and this involved a great deal of hard work, especially in clearing the saw dust and the unusually large amount of offcut wood. This was a major task in which Alan Williams was a willing helper, acting as labourer as well as supervisor. The sawing was completed in about 6 weeks by the end of November 1986. This included the logs which were to have been sawn at the Kingsland Sawmill, for after sawing eight larch it was unfortunately found that the facilities there were not really suitable for logs of the size and shape needed for *Spry*

No special arrangements were made for drying and seasoning the sawn planks, but the felling time of August–September, was reasonable favourable in terms of the sap flow in the living trees, which was beginning its Autumn easing off.

Most of the oak was stored outside on the Blists Hill Museum site, put 'in stick' with battens between the planks in the normal way, and was able to dry out slowly under generally favourable weather conditions, over the first Winter, through the following rather wet Summer and through a second Winter, before use. A small proportion of the planks were stored under cover for faster drying out and early use, but were not in quite the same condition as those stored outside.

The drying out of the whole Douglas fir trunks, felled later and required for the masts and spars, was more of a problem. Inevitably in uncut trunks, the drying out takes place from the outside diameter in. The shrinkage of the drying outer layer onto the still wet inner core causes stresses and radial splitting. To minimise this the trees were de-barked and painted all over, including the end grain, with two coats of a wax suspension to retard surface drying and this was very successful.

CHAPTER 5

The Rebuilding Work

Planning and preparations

Concurrently with the estimation of the timber requirements, the procurement of which was described in the previous chapter, in early 1986 Alan Williams was also planning and organising the provision of the facilities needed to carry out the reconstruction of the vessel. This was part of his remit to carry out a detailed feasibility study and costing for the project which was his first task upon joining the USNT Staff as Master Shipwright, in March 1986.

Three years had elapsed since *Spry* had been brought to Blists Hill while further funding was being sought and a suitable Master Shipwright appointed to enable the rebuilding to proceed, during which time a large Government surplus corrugated sheet steel hut, of semi circular end elevation, (Figure 23) had been procured and erected over the vessel to give weather protection to it on its hard standing. This came from Beavers Ltd., at Lea Brockhurst. This improvised 'boat shed' required raising by 5 feet to provide room for working around and over the hull. Later, a second hand prefabricated unit, originally a deep litter chicken shed, was obtained for use as a workshop. Later still, after Alan Williams had arrived, an extension was added to the 'boat shed', thus enabling the prefabricated unit to be converted into a much needed Tea Room for the visiting public, adjacent to the boat shed. The workshop, now in the main shed extension, was provided with electrical and plumbing services. Basic machinery, namely a circular saw, a band saw and also a planer-thicknesser, required procurement and installation. Work benches, a lock-up tool store and a small office and messroom were also needed and were provided in the shed.

These were the first minimum essential preliminaries in Alan Williams's programme, finalised in April 1986, covering an estimated three-year period to December 1989 for the complete rebuilding of the hull.

In this it was envisaged that the basic preparations would take 6 months, with work actually starting on the vessel in December 1986. The re-framing was to be completed by September 1987, the re-planking by December 1988 and the final work in replacing the hatch coamings to complete the hull, by December 1989.

The plans, which were presented in the form of a detailed proposal and a bar chart, were approved. Under the general supervision of Mike

Figure 23: The 'Boat Shed' at the Ironbridge Gorge Museum's Blists Hill open air museum site.

Day, the Curator of Blists Hill Museum at that time until the end of 1986, Alan Williams was given total responsibility, without technical or curatorial interference, including the authority for direct procurement of material and services. For this a special petty cash float was provided for him to purchase small tools and consumable items. Also, open orders to agreed maximum values were established for regular servicing requirements such as tool sharpening. This arrangement was essential because of the specialised nature of the work, in relation to the other activities on the Museum site.

The first phase of preparation, the improvements to the boat shed facilities, were completed somewhat behind schedule, in December 1986, with mounting criticism and impatience over the length of time elapsing without work actually starting on the boat. However it was essential to complete the preliminaries first. The delay was due to the progress being considerably hampered by shortage of labour, the only assistance being initially of a casual nature from other areas in the Ironbridge Gorge Museum. This in turn was caused by the expected help from the Government aided Community Services Programme, not being available until November 1986 instead of May/June 1986. This was very frustrating to all concerned in the enthusiasm to make progress after such a protracted period of hiatus since *Spry* came to Blists Hill.

After these initial delays, work proceeded at a satisfactory rate, although there was never more than 40% of the planned manpower available. This was due to the skills of the shipwrights and the efficiency of application of the planning, plus the considerable efforts of the available labour. Phase 1 was therefore eventually completed on time and within the cost budget.

The rebuilding work

The following detailed account of the stages of rebuilding is substantially in Master Shipwright Alan Williams own words from a series of recorded interviews with the Author. (Figure 24) They are reproduced in this spontaneous form for added interest and colour.

The account of the rebuilding work by Master Shipwright Alan Williams

Building a team

We had a community programme written for the project, getting folk to come in. They were from all walks. Some had never worked before in their lives, though they were in their early twenties, being unemployed since they left school. Some had been made redundant from previous walks of life, some had taken voluntary or early retirement and all these had to be brought together into a working homogenous team. Those who had never worked before didn't even know how to get themselves up in a morning with a packed lunch and into work at eight thirty in a morning so this was a personal disability and was the biggest step for them. We had to teach them all the basics – how to lift heavy objects, how to use a sledge hammer, how to work together as a team, as a unit and have an identity. Seemingly that is what was missing whilst they were unemployed.

It was planned that we would have myself, one carpenter and four labourers. The four labourers were divided, each labourer would do three days per week. So we'd have two carpenters, including myself, and two labourers each day, except for the double day – when we would have two carpenters and four labourers, so we'd plan our major moves of material or stock or major works for those double days. The arrangement worked out that we had a carpenter and a labourer, or 'offsider', so the carpenter could keep ploughing ahead doing the actual carpentry side. The offsider is the 'side kick', a 'gofer' – the carpenter's mate for everything that needs doing – going to make the tea at ten o'clock or getting more nails from the store. To create this fit, if you imagine that they came out of the unemployment 'swamp', we had to create an identity or a sense of belonging, for there were some two hundred other community programme workers, all on the same site, all out of the same 'swamp' and all lacking this identity. To create our own identity we'd

Figure 24: Master Shipwright Alan Williams.

buy different coloured overalls, we'd buy better quality work boots so that we stood out from the rest of the gangs on site. We had different tea break and lunch times, again to denote the difference. Then again between the gangs. We had two gangs, as each carpenter or each boatbuilder had a gang in its own right with his offsider, so you could start creating competition between port and starboard. You'd get one gang working down one side of the boat and one gang working down the other and by a little bit of stirring, and encouragement, prompting and ribbing, they would rise to the occasion. Mostly they became so keen that even on their days off, the blokes would come in, knowing full well that if they showed their face they would get roped into doing something, and they were quite keen to do this.

We did have a fair turnover of blokes, how much it was because we drove them too hard or how much because they couldn't fit into our regime, or because they got other work more in line with what they wanted to do, I don't know. I hope it wasn't the fact that I drove them too hard. We had a lot of people put upon us because the programme was designed to get people off the unemployment list, so they got put on the Community Programme scheme whether they really liked it or not, and for what was only marginally more money than they would get on the dole. So some of them really said, while we're going these lengths we might as well get a proper job. We lost quite a few of them that way.

With regard to how long it took to make people into useful workers, and to estimate how long jobs would take, I calculated the work, originally, on what the carpenters could do. The labourers were not calculated in as progressive man hours, they were just supportive and if any of them showed any aptitude for carpentry or taking initiative into other areas, that was a bonus for the project. I really based the whole programme on parameters of the leaders of the gangs.

We did try to teach those that showed some aptitude the basics of the methods and the use of tools. A lot of them were like a blank page on which you could teach them or instruct them and they would do as you would tell them. Those who had been in work before had their own ideas, most of them were on the right lines some of them had bad ideas to change, like how to lift things. You can't afford to lift things with a bent back, especially the weights we're talking about here. Then when it came to the tools like using an adze, bandsaws, planers, circular saws etc., we showed them the basic usage of them, the danger areas or the pitfalls, the things to avoid, and supervised them on the first half dozen applications that they would have to do it, and if we thought they were capable, then we would let them be. One or two of them didn't seem capable and therefore we just put them on other projects. They were a bit bitter at not being allowed to do as others did, but the job still had to go ahead and I couldn't really warrant spending the time to teach them the finer points. I really had to protect them from themselves from a safety point of view.

I got a lot of flak from the Community Programme director or manager, over my safety philosophy in that I wouldn't nurse people, I wouldn't cosset them. If I saw someone doing something dangerous I would stop them. When they were being told about the machines, I would say 'there is the safety gear, the ear muffs, helmets, gloves, goggles etc., for your use. If I see you doing something particular where I think you ought to be using more protection, then I'll suggest it'.

Most of the blokes wanted to join in doing the work, they didn't want to be just offsiders or 'gofers', only one or two of them did, but most wanted to play a contributive part which was very refreshing.

The equipment used

With regard to the actual facilities to carry out the work, we started with a green field really here, so not only did we have to build all the resources for the boat, we had no equipment and we had no forklifts. The building wouldn't accommodate fork lifts. We had no gantries or overhead cranes. We had to do it all by hand as the Egyptians did when they built the pyramids. We used crowbars and fulcrums, rollers, blocks and tackles, ropes, trolleys – we had a couple of two-wheel trolleys, and in addition to those we had sheer legs and we had a chain block. This damn thing was as heavy as what it could lift, just about. The shear legs were made out of old telegraph poles, lashed together at the top. But with those very, very basic items there's nothing you can't do. The Egyptians built pyramids, the Ancient Britons built Stonehenge. Boat building hasn't changed in a thousand years. Granted you have power tools now, you have timber arriving on site pre-cut to at least a near specification anyhow, sided at 4 or 8 inches. But thereon, boat building hasn't changed really at all. Somebody's still got to make a spile, which is a template. Somebody's still got to take this spile and select a piece of wood to cut to it. The tools available nowadays do save some of the back-breaking work and time in sawing and planing and so on.

We did also have a chain saw. Some of these slabs for cutting frames out of were too heavy to lift and put up onto the bandsaw. Well, they all were. So what we did was to chain saw them out of the slabs lying flat on the floor and we allowed roughly ¼ inch for planing off and cleaning up. In this case, it took almost as long to clean the roughed out frame up as I would think it would have taken had we done it by hand using a pit saw albeit with more physical effort. Had we got the slabs of timber over a pit, and a chap down below and one on top, sawing to the line and to the right bevel first off, I think we could have done it almost as quickly. I'm not sure, but it would an interesting exercise. But no, at the end of the day, somebody's still got to grunt and lift, push and shove.

Shaping the timbers

What we had to as the very first thing before any reconstruction work could be tackled, was to pull the vessel back into shape, as it had sagged open and distorted due to the lack of support and the weakened frame fastenings. This had got worse during the drying out in the three years *Spry* had been standing at Blists Hill. To get it back into shape, we made a complete external cradle of moulds to constrain the hull and let us true up the timbers to their original lines. I did this by judgement of eye. This provided the necessary accurate basis for us to go ahead and shape the new timbers to replace the old ones.

To give an idea what the moulds were like, imagine a wide flat capital H with the corners scrolled in and these were made of 14 × 6 inches baulks of timber all braced and jointed so that they actually had quite a bit of stiffness in their own right, but to stop the top edges spreading we put wires and bottle screws across and set those up in tension.

With regard to the actual rough shapes, before we could really get the shapes, we had to take a spile, which as I said before, is a template. Most people think of a template today as being something that fits exactly into the position that it's meant to represent. But to do that on every frame, bearing in mind that each frame consists of eight parts, would be very laborious and no two parts could be used consecutively on the next frame. So what we do is, we use what we call a spile which is a thin piece of plank or plywood or whatever, hold it up in situ, and you take a parallel mark with a chock of wood from the face of what you're trying to mimic. You mark that onto the actual spile, around the curve with the angles and bevels. Then you take the spile and offer it up around a selection of slabs, trying to find one which will match it best. Having done that, you then transpose the marks from the spile with the same piece of wood, back onto the selected piece of timber. You join up all the dots and then cut the shape out with the chain saw.

Having then lightened the piece of timber – the slab might have been a ton before we took the chainsaw to it – now having a piece of timber which weighs 100–150 lbs perhaps, we could take it onto a bench or put it on trestles and clean it up to the actual line, taking off the roughly ¼ inch we had left on in chain sawing. The greatest boon to boat building has been the electric plane and with that we can clean up all the outside curves. Unfortunately they don't make a compass plane or a plane which you can plane concave surfaces with, so all that has to be done with the adze. All credit to these lads we had, we taught them all, to a greater or lesser extent, to use an adze. I had a selection of about six adzes and each lad had one for himself for the duration of his stay with the project, for him to look after, to maintain. Every adze is different, so if you gave a lad a different adze every day, he'd never get proficient. So you see, they all contributed to shaping the frames.

Replacing the frames

The frames then had to be fitted, one at a time, in place of the old rotten frames, within the old external planks, which were still left in position. Having got the new frames all cleaned up with their respective bevels, they would actually go in there pretty much as first time fits.

Going back quite a bit to the actual fundamental decision-making time on the boat, the decision to replace 100% of the frames came about because the boat is planked internally and externally. We could never access the frames in future had we left half a dozen which were kind of all right at the time. They would be the first to go in the future, so we couldn't afford the possibility of having to replace them in say ten years time. You would need to either rip the outside of the boat off, or the inside. So we had to replace them 100%. So that means each frame – although it wasn't totally rotten – we'd bash and chop and split out the old frame and try not to damage the outside or external planking. Don't forget, there were hundreds of nails in the old frame. She'd been fastened originally with – remembering that each frame consists of two girths – with one nail in each girth, so you'd got two nails per frame. Now she had been re-fastened later, that is she'd had another set of nails per frame and she'd also been repaired. That means yet another set of nails per frame. So where originally there had been just two nails, in many areas there were four, six and even eight nails through each plank at each frame. These are nails through the outside planks into the frames. We must also remember that the ceiling, which is on the inside of the frame, is also nailed to the frame, so the number of nails into each frame altogether is double what I have just said. Where I said two, that should read four, six should read twelve and eight, sixteen. So perhaps you've got sixteen nails per foot run of frame. Now to find somewhere to get a chain saw through the frame, and split out the frame without damaging the saw, or one or two chisels, was an absolute nightmare. There was a real forest of nails. We could split off the top 1½–2 inches of frame and hopefully expose all the nail points and then we could drive them out backwards, then get the frame out that way. But the frames would never come out in anything greater than 12-inch lengths. Having taken the spile to the frame, made up our new section of frame and we'd got it in place, we would then nail all three pieces together. The way the gangs would work, we had Hugh who was a boat builder and who was my apprentice in Australia – he came over for six months – his job was to replace the 'floors'. They're the bottom two timbers of the frames lying transverse to the bottom of the keel. So he'd replaced the two floors and then nail them together. Then he'd move down to the next frame and get it out. Then Paul – we had two Pauls – Paul, who was a skilled carpenter, and we had another lad who was called Paul, Paul Turner, whose father had a boatyard on the canal, less skilled at that time but who showed great aptitude, so he led the starboard gang. So we'd got Paul the tradesman carpenter leading the port team and the other Paul leading the starboard gang and they would would fit the first 'futtock' and then the second 'futtock' – these are the middle frame members above the floor – and then they would fit

Figure 25: Framing nearly complete, showing old planking retained temporarily in position.

the top timber and nail them all together. That would be a complete frame. By that stage Hugh, who'd been on the next frame, had finished the floors and moved on to the next so we had no overlapping. We temporarily re-fastened the new frames to the old external planks to hold them in position. All this was done within the external cradle we made originally to hold the hull in shape, to hold the whole lot together and to stop the frames falling over. We coach screwed the old planks onto the new frames, (Figure 25) and this was beginning to put some rigidity back into the hull structure.

Hugh, Paul and Paul number two worked along the body of the boat in this way. Myself, what I was doing first, was making a set of drawings for the boat – line drawings and structural drawings. Then my job was to start with the framing from the back of the boat and work forward and meet them. My main problem was that I hadn't got a piece of timber to do the stern or the aft dead wood, and getting this piece of timber had been bugging me for months and months. In the original acquisition of timber we hadn't been able to identify a suitable piece and now here we were, we'd got three gangs working back towards me like nobody's business and I still hadn't got the back sorted out. This piece of timber, to be in oak, was like a large boomerang or like a large shelf bracket, about 9 ft on each arm, and through the corner we needed three or four feet of timber, so you can imagine it's quite an extreme piece of tree. In desperation I phoned a friend of mine, Ken Benbow, and said, 'look we really are crying out for this tree, can you help?' So he said, 'Yes, come on over.' This was about ten o'clock on a Tuesday, we went round to his place and we drove around the country – actually most of the county – looking at all the oak trees from the car. We didn't have any specifically in mind. Having seen a likely one eventually and walked across to it in the middle of a field – a hay field

Figure 26: External shaping in the rear frames and showing new keel in position.

I think it was, and they had just taken a cut off it – we decided that was the one we would go for. It was actually on Sir John Corbet's land, one of his fields as luck would have it, for Sir John was on the USNT Committee and in fact the original Chairman. Ken Benbow phoned up and got permission to cut the oak down that evening. Wednesday morning I went out there and joined the timber feller and he felled the tree, trimming off all the excess material and actually running his chainsaw down the sides of the tree to make me a 10-inch wide flitch of oak with these two arms hanging out 10 feet each way. Wednesday evening we arranged transport. Thursday morning the transport picked it up and dropped it here at Blists Hill just before lunch time. The following Wednesday or Thursday the piece of timber was all shaped up with all its housings and bevels and ready for fitting. So the timber had been shaped from the tree still standing just a week previously. The only condoning factor was that the tree was actually standing dead – a lot of the moisture had gone out of it already but it was still fairly green. That was the biggest and the most 'panicky' piece of wood we had to do. Of course, having got that stood up in place, we could then start to put the stern post up. The transom I made on the floor of the workshop – it's much easier to work flat than working 8 feet in the air. It's also made of oak, and I nailed it all together and then lifted it up in one piece. Again, this was heavy stuff, the deadwood weighed about a third of a ton, maybe a bit more. The transom weighed a quarter of a ton. For lifting these slabs of wood around, again we had the sheer legs, the two telegraph poles, and a chain block. It doesn't take long, it takes a couple of hours to set up the shear legs, and an hour to lift the piece of wood in place. Had we gone and got a crane, which seems to be the usual philosophy, you'd have a minimum hire of 6 hours, you'd have to take the roof off the shed in the area we wanted to lift. There'd

Figure 27: Visit of HRH the Prince of Wales, 6th March 1988. The Prince accompanied by Mr. David Lawrance, Chairman of the USNT, and Alan Williams.

be communication problems – you wouldn't have eye contact with the crane driver – good heavens, it's a hell of a lot cheaper to use the shear legs, primitive though they are.

Once we had the stern assembly in position we could start replacing the frames in the stern. The frames at the bow and at the stern are called cant frames. They are still double sawn frames but they don't cross the centre line because you have the forward dead wood and the aft dead wood in the way. They actually abut the deadwood in shallow housings. These aft frames have the most extreme shapes of all the frames in the boat. They have some of the most confusing and convoluted angles on their faces and proved great fun to fashion and make up. I'd actually bet the gangs working from forward aft that they wouldn't finish their framing before I finished the stern framing. They got to frame 32 or 33 just as I was finishing frame 35, so you know, I'd only just got my part of the deal done by the time they came up and met me. It was a close thing!

As a point of interest, cant frames line up differently to the centreline than the others. All your normal frames run square to the centreline, but the cant frames usually run more square to the external planking.

Imagine the bow coming into the stem, you need to have your frames running more square off the planking, otherwise you're going to have a very squashed rhomboid of a shape if you try to keep square to the centreline. The cant frames on the *Spry* though, were not too far off square to the centre line both fore and aft.

So this was a major job done, the replacement of all the 40 frames, including the stem and stern posts. (Figure 26) We had started this in January 1987 and we finished in February 1988. Just after that a very big event took place. HRH the Prince of Wales inspected the vessel, during his visit to Blists Hill on 6th March 1988, to open the Wrought Iron Works. (Figure 27) The Prince seemed very impressed with what we were doing and the size and nature of the project.

Commencing the planking

Having finished all the framing now, the outside moulds began to be a hindrance. We couldn't take them away entirely as the boat still didn't have enough integral strength to support itself, so we cut off about 2 feet to 2 feet 6 inches from the outside moulds thus exposing the top couple of feet of frame. The idea being, we could wrap a couple of planks round the outside of the boat and one on the inside to tie the tops of the frames together – so having lowered the moulds we also lowered the wires running across the tops of the moulds to hold the boat together, to stop it spreading.

We then needed to tie the tops of the frames together to give rigidity to the structure. The top outside plank is the sheerstrake. That's thicker than the other planking – it measures approximately 3 inches thick. Its width varies from 10 inches down to 8 inches. The plank below that, the top strake, is back down to 2 inches thick. So the first thing to do was we faired in all the tops of the frames, established our sheer or line of plank to cut the frames off to. We put the top planks on and nailed them and then we put the sheer strakes around. This was very difficult. The top planks we steamed in the steam box and then wrapped them around. We couldn't do that when we came to the sheerstrakes, as the darn things were 42 feet long and 3 inches thick and weighed some 400 lbs say. Had we steamed the ends of them, it would have taken us 15 minutes or so to get the timber up in situ, and by this time of course, you would've lost all the benefits of steaming. So we decided to wrap them round cold. To do this we got the front ends entered into the stem and fastened and clamped them, taking every precaution we could. We even put ropes around to stop them suddenly springing out had the clamps and the nails failed. With the front ends entered, the after ends were sprung about 8–10 feet wide of the boat so we had those supported on a sort of runway leading up to their final resting place. Between the two ends of the sheerstrake we had a big five fold block and tackle. Over the next hour or so we just gradually hauled on the block and tackle until they were closed. As each plank came around each frame that it contacted with, we'd quickly throw a clamp on just to ease the tension in bringing it to the next frame, lessening the chance of it breaking. It was actually very hair raising, because all the staging was set up around the outside of the boat and to put a clamp on, you had to walk around the outside of this straining, groaning, creaking lump of timber which at any moment could actually blow apart. So we breathed one hell of a sigh of relief, as you can imagine, by the time we got the two ends in. Before we could do this, we had to actually shore the frames apart. You can imagine the crushing force of wrapping these two planks around – we put 4 or 5 Acrow props across the inside of the boat and set up quite a bit of pressure in it to counteract the crushing force from outside. Fortunately all went well!

The stern wasn't so bad because it's got a much gentler shape. It's quite a mild shape really. What we have got there is twist. Now we couldn't do that cold. It was possible to steam the after ones, because they were much shorter and not quite so wide, and we needed to because of this twist. Having done the outside two planks we then had to fit the beam shelf inside and this caused me hours and hours of wondering how the blazes we could do it. Bending things round an outside curve is relatively easy, for as you come against each frame it's an automatic stopping place to lever against. If you've got planks you are trying to wrap round into the inside of a shape, there is nothing the action can come up against to lever around. So with the fore end of the beam shelf touching and the aft end touching, the middle of the shelf was some four or five feet away from where it actually needed to end up.

We had long clamps, with threaded bars and we started with these, cranking and bringing it in but it was making most unhealthy noises and so we decided to call it quits there. We then rigged up the steam pipe out of the steam box underneath the beam shelf and covered the whole thing over with sacking and polythene to try and trap as much heat as we could. We fired up the boiler for two or three hours and as the temperature built up and the timber soaked longer and longer, we gradually worked the clamps in as best as we could. Trying to work blind – with gloves on of course, underneath the canvas, in all the scalding hot steam, was quite a problem, but it came in gradually and we eventually got it right in without any trouble at all. Having fixed one, and got the know-how, the second was straight forward. It took about a day per beam shelf, to get them in place, but having got them there, we could then rivet the whole lot together, through the sheer-

Figure 28: The for'd deck beams in position.

strake, through the top plank, through the frame and into the beam shelf. I used the word rivet, I should have used bolt as the heads are really peened over and not riveted over. Riveting over implies quite a large bulky head. Peening over is really just a flaring out of the end of the nail. We did the peening cold. For a dolly we used our anvil. Because we're talking of such large sections of iron – they were ⅝-inch square bar I think – we wanted quite a massive dolly on the outside. We thought the anvil would be the ideal job. So we made a 'sky hook' that we could hang from the actual framework of the shed and we put the block and tackle on that and hoisted up the anvil. I don't know what weight it is, it must be two hundred plus pounds. We poked the corner of the anvil into the countersunk hole and then riveted against that. This caused great fun.

The decking supports

Having got the beam shelves wrapped around inside the frames, we could then put in the deck beams. The boat's only got a bit of decking forward and a bit aft. It's about sixteen foot each end of the boat that's decked and that consists of . . . one, two, three . . . actually six deck beams each end. These are 6 × 4 inches oak with a 4½-inch camber cut in them to shed the water. Between the deck beams there are the carlings. Carlings are frames on which you mount future fittings so at the same stage that you put your deck beams in, to stop the deck beams moving around, you also put the carlings in. At this stage they act like spacers but they are not designed as spacers, they are designed as frames on which you screw or mount future fittings. So you have two large carlings for the legs for the anchor winch to go through, two carlings to take the samson post, two more carlings for the crew accommodation scuttle and two large carlings for the cargo winch. These were all put in and fastened. (Figure 28) The deck beams were carved with their 'Certified to accommodate the Master' in the rear cabin area and 'Certified to accommodate four crew' up for'd.

Completing the planking

Having got those in place and fastened, we felt quite secure, actually secure enough to be able to take the remaining outside moulds off the boat, so we could get round to the bilge and replace about three planks outside the boat. Also inside the boat, the stringer – the bilge stringer – which consists of two pieces, one each side, and each approximately 40 feet long – one piece is 5 × 8 inches, the other is 5 × 5 inches. These are really like longitudinal stiffeners. On the outside of the boat, on top of the planking, you then have the bilge keels. So before we could start all this, we took down the outside moulds, and we took off 4 of the old planks from the turn of the bilge right from one end of the boat to the other, faired in the frames in that area and planked up. These planks, two of them were full length strakes, and two of them were stealers. Stealers are planks which don't run the entire length of the boat, as opposed to a strake which is a plank which runs – not necessarily in one piece – from end to end of the vessel. So we put on these two planks and two stealers. One stealer ran from forward, finishing on the aft quarter, and the other one running from aft, running forwards finishing on the bow forward quarter.

We started the planking at the turn of the bilge rather than following a more usual sequence, starting from the keel, because our problem right from the word go, was that the actual structure of the vessel was never strong enough or sound enough to be able to replace timber in the conventional order. If you were going to replace a keel you would normally jack the boat up and put a keel underneath, but we couldn't jack the boat up at this stage because the hull wasn't strong enough to bear the strain. When we had virtually re-framed the boat, we then got rigidity at the top of the hull having tied up the tops of the frames together with your beam shelf and your sheerstrake. We still couldn't jack the boat up even at this stage though, because it still hadn't got the integral strength down below, so we came down to the corner of the bilge and replaced four planks in the bilge and the bilge stringer inside. We also put the bilge keels on. Incidentally the bilge keels, unlike the main keel, don't actually penetrate the hull, they are only fastened on the outside as additions. So they are not really a structural part of the boat. They are riveted through, but they don't actually penetrate the hull. The main keel actually mates on the outside face of the frames whereas the bilge keels are just slapped on the outside of the planking and then riveted through. However having got the bilge keels on, this did stiffen up the whole corners of the hull and we could then jack the boat up and support it on the bilge keels.

Replacing the keel and keelson

We built a gallows fore and aft underneath the sheerstrake forward and the sheerstrake aft to support the overhangs and then we could knock out all the blocks under the central line – and the keel virtually fell off – it took us just an hour. Within a couple of hours – we had already prepared the new keel with its attendant rabbet and scarf – we had it up, all in the same day. So what is usually a mammoth task only took a day.

Having got the new keel up in place, as it was fastened by spiking through the frames into its top face, this had meant rolling aside the keelson to access directly on top of the keel. There is only one spike per frame, remembering that each frame consists of two sections, so when we came to put the keelson back on top of the frames, the bolt which goes right through the whole lot goes through the other frame section so as not to foul the keel spike. The keelson that was in the boat when we got it was some 16 inches square, 60 feet long and of Douglas fir. It had been replaced, we don't know when, but I would estimate, perhaps in the early 1930s, and it would have most likely have been at the same time when the boats sides were raised. The original one would obviously have been subject to a lot of wear through years of loading stone day in and day out and it could have been replaced because it was damaged or generally worn out. It probably would have been made from pitch pine. I'm sure this original one was not as large as 16 inches square – that seems quite excessive to me. The bottom two inches of the keelson were actually rotted where it had been lying against the frames, but otherwise it was in good condition to use again. I wanted to reduce its size to bring it into something like visual correctness and taking the rotten part off allowed me to do this nicely. So, turning it on edge, I took 2½ inches off the bottom and about 1¼ inch off both sides of the keelson. That's putting it square again to approximately 13½ inches square. The other thing they'd done when they replaced the keelson – they were obviously in a very confined slipway or dry dock – was they put the new bolts in by driving them down through from the top, through the keelson, through the frames, into the keel and then clenched over washers underneath the keel. You wouldn't do this if there were room underneath to swing a sledgehammer up to the bolts which are 30 inches long. The result was, with the way they did it, when the boat went aground, it just took the heads off the rivets and popped off the washers, so the keelson was, in effect, barely fastened when we came to remove it. So to put it back as it should have been done. I opted to drive the bolts up from underneath and then rivet them over on the inside of the keelson. We were restricted for clearance but luckily, the boat was some four feet off the ground so it allowed us about 12–18 inches of clearance to swing a sledgehammer to start with, which is not much, but as the fastening goes further and further in, of course you get a greater and greater swing at it, of which you need every inch, as the bolts took some force to drive them up. We pre-drilled the holes of course. The bolts were an inch square, so we drilled ⅞-inch diameter holes to give a definite driving fit. The corners do most of the holding as they have no where to go other than to force their way through the wood. As the bolt is driven in further and further, it sounds at a higher and higher pitch because as well as getting tighter in the hole, the actual distance of resonance gets shorter. Even sitting in the shed down here or whatever I was doing at the time, I could tell exactly how far the blokes had got in driving these things in by the sound of the hammer on the bolts. We would have four people on the job with two people driving and two people resting at any one time, to do it non stop, for the moment you stopped driving the bolt, the grain would start closing in and there'd be the devil's job to get it started again.

Actually, when the replacement keelson had been fastened in, they had countersunk the heads of the bolts some 1 inch to 1¼ inches into the keelson and this had mashed or broken the actual hole edges where our new rivets would come up through – so before we started driving in the bolts, we let in patches of oak into the top facing of the keelson to make the hole edges good. These are called 'Dutchmen' from the tale of the Dutchman putting his finger in the hole in the dyke. This is a sensible idea, making good a hole by letting in a piece of oak. I think they were about 1¼ inches deep and they gave nice firm and hard surfaces on which we put the washers and then riveted over. Then we could of course drive the bolts right up through as I have described. It took just over a week to drive in all these fasteners and to rivet them – myself and one other chap were on the riveting gang whilst the others were sledging the bolts up. We had an old Acrow-prop – when digging trenches you have to keep the sides of the trenches apart with very short Acrow-props – we got one of those and welded on the top of it something like a mortar launcher – just a bit of tube – into which we put a bit of 4-inch diameter iron shafting, and this was our dolly. We put a couple of circles of motor car tyre in the bottom of the tube, to take the shock of the impact out and this gave the dolly some bounce rather than it having a dead impact. You need some bounce back so it taps the rivet back in again and if there's no give with riveting on top, with the base of it sitting on concrete, the concrete can powder and become loose. You need some spring to hold it, so it makes sense. With the adjustments we could actually wind up the thread, as we had to force the actual dolly against the bolt head. We could then just drop a washer over the top, cut the bolt end off and peen it over. When it was all done, the whole backbone of the boat was back together again, as strong as she was a hundred years ago. It was a very satisfying thing, for up until this stage the boat had really been floating in mid air all the time in an unstable state – to have got the keel actually fastened in gave me great peace of mind.

The planking

Well, with the boat's main structures now all held together, we could concentrate on planking and really we'd just got to fill the gap between the keel and the bilge keel planking. Then from that it was just top sides planking which is from the bilge up to the sheer strake. The first plank where you come out from the keel, is known as the garboard. Originally this was elm, but we were putting larch planking back wherever we could. Putting the garboard in was quite a nightmare in that at the for'd end it goes from the vertical to a horizontal in about 6 feet. So that's one hell of a twist – it also goes round a curve as you're coming out of the bow. This caused me quite a few sleepless nights, so having shaped it to the spile, we threw the plank in a pond outside, just a few yards away, to soak overnight. Next morning we fired up the steam box and steamed this wet plank for a couple of hours. With all the jacks and props and wedges and clamps ready, we got the front end entered and with four chaps heaving round the after end of it, the darn thing went round remarkably easily. I'd been dreading this for weeks and months, if not even twelve months, having seen the problem way ahead of actually tackling it, but at the end of the day, it went round very very easily and it was a bit of an anti-climax.

The aft end of the garboard has got a big twist in it as well, but it's a more gradual twist, it does it in perhaps 12–15 feet from the vertical to the horizontal. The drawback there is that it's not going round a curve, if you are twisting something from vertical to horizontal round a curve you'd virtually just pull it around from station to station or from frame to frame. Because each frame has got that much extra twist than the one before, it naturally flows around. On the aft ends you have no such curve to wrap around, so we had to use a couple of bending bars, just a couple of bits of 4 × 2-inch wood, clamped at right-angles to the plank, with which we could actually twist, using them as levers. So having got the aft end entered in place, we then got the chaps hanging on these bars, levering and twisting it round to get it to the horizontal before we could even think about clamping to the frames. In fact the after parts proved more difficult but at the end of the day we got both of them in there which was very reassuring. Having done that, the rest of the planking, in theory, should be a piece of cake.

The procedure for fitting a plank is – we call it 'spiling'. I've mentioned using spiles or templates before, on the frames, but I'd like to describe this a bit more now we're on the planking. What you do is you have the parameters that you want your plank to fit, that is the widths, marked out on the frames. What we do then is we lay and twist a light plank which is the 'spile', maybe ¼ or ⅜ inch thick, onto the frames. The old planks couldn't be used as spiles, one – they were very scraggy and two, they had got their twist in them already and of course, they won't untwist to lay them on a flat plank. So we had to go back to first principles – taking a spile, which is the very thin plank, which is laid in the aperture where you want the plank to go. If it doesn't fit right out to the edges, we lay it in place and just tack it with nails – we're talking 30-foot lengths of planking here – so we might have to join two or three pieces of spile together. The spiles were actually just offcuts of larch from when we were cutting the larch planking. Because we wanted only 2-inch thick planking, the first 3 or 4 inches off the girth of the tree is no good to us because it's just too narrow and has too much sapwood, so this stuff had got to come down to ½ inch, I think it was, and we had a great pile of it, like larch-lap fencing. The other important thing is, when you wrap the spile around the boat, is not to put any unnatural stress in it – you can't edge bend, you mustn't try to force it edge-wise. It must lay naturally in the space you want it to go. If you try and force it, when you take it off the boat, it will take its natural shape again on the plank and it will not be representative of the true shape which you're wanting. It's very easy just to try and force it or to push it somewhere where it doesn't want to go and that usually winds up with you coming unstuck. Having got the spile in place on the boat, we would then get a batten or a chock. We used pieces which we actually used right through the whole project – they were bits of Formica about 4 × 6 inches – white Formica , it's nice and light to see where you are working, for it's a bit dark around the boat. We also had some 3-inch and some 2-inch battens if we were working on a very narrow plank. We would write on the spile which size batten we were using, so we could just go straight back to it.

The spile fits in the aperture, plus or minus an inch or so – it didn't fit to the edges – and then we took a parallel mark off the plank below and transferred that parallel mark onto the spile so that represented the lower edge and we did the same for the upper edge, laying the spile to our mark on the frame where we wanted it to finish, and then mark below that. We would do this on every frame so we'd have 20-inch centres for all our marks. Also on the stem, if the plank was coming to the stem, we'd just lay the spile up against the stem and take a parallel mark and that gives us the angle of the plank at the end. Each of these marks would be numbered 1 to – depending how far the planks were extended down the boat – say 10, 15 or 20. Then we'd have to take a series of bevels because the boat's like a barrel. You've got planks fitting round a curve, so the edges of the plank don't want to be square – they need to be slightly over-square. This bevel changes along the full length of the plank. Right in the stem, it's pretty straight so you've got virtually no bevel at all. When you come to the turn of the bilge where it's tightest, you've got perhaps 3/16 inches of bevel over the 2-inch thickness, which is quite a bit and if you don't put it in there, well you just end up with acres and acres of caulking. Because the bevel changes along the length you need to take some twenty bevel measurements along it. Actually we only bevelled the bottom edge of each plank, the top edge we left square and we put all the bevel on the next plank up.

Having taken all our parallel marks from the sides of the planks and marked them on our spile, having numbered the positions representing each frame and taking a series of bevels, we could then take our spile off the boat, and offer it up onto a plank. We had to go searching through our pile of planks out in the yard to find a plank which would represent something near to the shape of the plank we actually wanted to finish up with. The boat is a three-dimensional shape, and although the planks can be bent in two dimensions – they're part of a flat sheet and you can't wrap a flat sheet round a three-dimensional shape. You can put a bend and twist in or you can put a bend and bend in, but you can't put bend, twist and bend in. One of the dimensions you have to actually cut into the plank, so we cut all the edge bends or the edge shape in the plank, which just meant choosing one with enough curve to cover the shape. The round-the-boat bend and the twist – those get put in when we steam and bend it round the boat. So none of these planks actually runs straight, some of them are a great elongated letter 'S' shape, which is hard to believe and it runs totally contrary to what you might imagine. If you are looking at a boat and are asked to guess which way a particular plank curves, it will probably curve in the opposite way to that you might think.

The trees we selected for the planking were not necessarily straight trees. Larch in particular, is very useful as it usually has a swept butt, the bottom 6 to 8 feet of tree has quite a sweep in it. So having selected our plank from the yard we would bring it into the workshop, lay the spile out on top of the plank and tack it in place so it doesn't move. Then transfer all our parallel marks with the same chock that we used to mark them initially, onto the plank and number each station yet again on the plank. We could then lift off the spile and put it to one side, that's then finished with. We then had to think how much the plank is going to increase in size to the outside, because what we'd actually taken is a spile of the inside face of the plank. Because we are going round a barrel curve, the outside face of the plank would be wider and we would then reckon that we could gain ⅛ or ¼ inch at the outside of the boat for the required bevel. We would set the saw up with a bevel on it at a slight angle, cut round the plank leaving the meat on, which would then be planed off to the series of bevels for each station we came to and fairing in between the stations to get a fair line. Having done that, then we could start thinking about a caulking groove. If we put the boat together just with these bevels we'd taken, in theory the boat shouldn't need caulking – each plank would fit edge to edge with no gap whatsoever. I can't claim to be that good a boat builder to guarantee water tightness and nor can anyone else. So all working boats were designed to have a caulking groove and the caulking groove only runs for half the thickness of the plank. So if you've got a 2-inch thick plank, you've got a 1-inch caulking groove. We aim to get about 3⁄16 inch of caulking width and that tapers to zero at the middle of the plank. So the first inside inch of the plank we aim to have an interference fit. The outside inch of the plank is a tapering aperture. Our measurements were quite accurate – the planks are heavy, they are awkward, you don't want to lift them around more than you have to, so the aim of the game is to have everything a first time fit – especially when you put them in the steam box. You don't really want to take them off and start trying fitting them. They pretty well all went round first time – I think there were one or two perhaps, no more than that, that we weren't satisfied with and actually had to lift off when they had gone cold and make slight adjustments to. They mostly fitted to within plus or minus 1⁄16 inch, which I think is pretty damn good. This has to be done for every plank. There are 13 strakes a side and two stealers. If we say there are 15 strakes a side with the stealers and each strake is made up of 3 planks, that's 45 planks a side – 90 planks in all. Each plank has to have this process and that's just on the outside hull planking. Then you've got the same number of planks on the inside of the boat and you do exactly the same process again there. So you are talking about getting on towards 200 times you go through this procedure, absorbing all the information, to carry it in one's mind, where you've go to, and interpreting shape and tolerances even though you've marked it all out. All is laid off onto the plank and all cut out. There might be things which look to your eye not quite right so you have to think 'do I try taking it off to correct it or do I leave it there'? — I really detest planking in a way, as it's so repetitively boring. You know you come in next morning and say, 'what am I doing today? Oh yet another plank!' It's not very conducive to keeping up your enthusiasm. All this took a long time. I can't remember when we actually finished – I think it took about a year to plank the whole boat inside and out.

I omitted to explain that because these planks are going round an outside of a curve, the inside faces of most of the planks needed hollowing out to fit the curve of the frames and this was also a variable. At the end of the boat at the stem the frames are pretty flat and in the turn of the bilge you might be taking out $\frac{3}{16}$, $\frac{1}{4}$–$\frac{3}{8}$ inches even – out of the inside face of the plank and of course you can't do that with an electric plane. That all has to be taken out with a hollow plane – a round bottom plane – I've got a couple of old wooden smoothing planes and we just planed up a radius onto the sole of the plane. Even though the planks were 12 inches wide it wasn't possible to bend them around the curve of the frames. They were already being bent in two dimensions and you'd be trying to bend them in another dimension. Another thing is, being 2 inches thick, they would crack along the outside face. You couldn't steam them, you had to do it with a smoothing plane by hand and this would take a lot of time again. We made jigs up to measure the actual curves to transfer them the full length. Very time consuming.

Steaming the planks

With regards to steaming, we try not to steam planks if we can help it, if we think they'll go round cold, we put them round cold. Planks will go round one dimension cold – no problem. So on the top side planking, for'd especially, where you're just going round the one curve, they go round cold quite easily. But twisting – they don't like twisting cold, so really in the aft end where you've got twisting in particular, then we'd steam those. For'd lower down, we'd steam those too, to get them to go round the sharp curve.

When you steam a plank, the plank loses some of its durability. I don't know why this is but my theory is that the steam induces heat and moisture which ruptures the 'valves' in the tracheid, between the cells in the actual timber structure. The same thing happens when you kiln dry timber. If you've got timber which has got a high moisture content, as all timber has when it's felled – of the order of 50–60% perhaps when it's green – in hard woods in particular – if you then whack them in a kiln, whether its a heat and steam one, or a vacuum and steam one, it pulls the water out faster than – I forget what the 'valves' are called in the cells of the timber, in the tracheid – than these 'valves 'can pass. They're like trap doors and the heating blasts the 'doors' open and they won't shut afterwards. In air drying timber, these 'trap-doors' open just enough to let the sap out slowly and then they are not damaged and can close again. Kiln dried timber is really like a vast drinking straw, you could stick one end in a bucket and could suck on the other. There is an easy avenue for moisture – and for rot spores – to penetrate the timber. Therefore kiln dried timber is an absolute No! No! in boat building. Steaming a plank, I feel, has the same effect in a way, though the timber you are steaming is dry you are then introducing steam and moisture and I'm sure the cells are stressed. Although they are devoid of any moisture to start with, they must have got air in them, so you are pulling the air out or you are expanding the air and causing damage to the cells. I don't really know the full technicalities of this, it's just a thought from my experience. So we try not to steam them if we can help it.

However we did have a lot of steaming to do on this boat. On many other jobs I've worked on before, we've only been replacing half a dozen or so planks, putting in a few here or there, and we'd end up boiling the planks. We'd be on the coast so we'd use salt water which is slightly better than fresh. We'd have a large 2-foot diameter pipe plugged off one end – dig a big pit and build a fire in the pit and poke the end of the pipe in the fire. Our plank would then subsequently be immersed in the boiling water in the pipe. It's a bit limiting because you can only steam a length of timber up to the length of pipe that you've got and also it takes a long time to get the water in the pipe up to boiling temperature. If you can get a tube boiler, a water tube boiler or a water jacket boiler then of course the temperature comes up much quicker and you don't burn as much wood or coal. You get the job done that much quicker.

I decided that here, we would steam the planks rather than boil them. I found an old cast-iron boiler up in the top of the Museum, which was an old central heating boiler out of a school. This was a cast-iron thing with a water jacket and a large enough aperture to the fire box to take pieces of wood. We'd got tons of wood offcuts from the edge of the planks which we could burn, so we didn't have to buy coal. There's a saying that you burn a boat to build a boat! So we had the boiler and we had the fuel to fire it with. We just made up a 17–18-foot long wooden box as a steam box with a lid on it to steam the planks and we took a 2-inch diameter pipe off the boiler. The boiler was meant to work totally full of water and was never intended to be boiled so as we wanted to create steam, we had a header tank which we kept topped up to a level such that there was maybe 4–6 inches of air in the top of the boiler from which we could generate steam. We ran a length of 2-inch bore pipe off the top of the boiler and introduced it to the steam box into two ¾-inch bits of conduit running along the inside of the box. These were perforated at intervals so we could introduce the steam right down along the full length of the box rather than just introducing it at one end. The idea is you steam a plank for an hour for each inch thickness. So if you've got 2-inch planking you

Figure 29: Removing a plank from the steaming box.

do 2 hours steaming. That's a general rule of thumb, but if you'd got a plank which had got a particularly acute bend or curve – then you'd give it a bit longer. If it's quite a mild shape then you could get away with a bit less. The garboard was quite an acute shape so before we put it in the steam box, we chucked it in a pond nearby overnight so it could take up some moisture, rather than trying to put all the moisture in by steaming and then we steamed it through.

Before we put a plank in the steam box, we usually have a dummy run to where the plank has to go on the boat, so all the blokes in the gang would know where they've got to be, what they've got to be doing, that we'd got all the shores and jacks and wedges and clamps set up and all the threads backed off, sledgehammers stood ready and so on, because once the plank comes out of the steam box, (Figure 29) the sooner you get it wrapped round the boat the less it's going to 'fight' you. It doesn't hold the heat at all long. You might be talking nearly 100 degrees in the box, and I suppose if you could make the box carry a small degree of pressure, then you'd get the heat up a bit higher, but the temperature drops off incredibly fast. You really want it wrapped round and held in place within 5 minutes. If you get it done quicker than that, then of course life's a bit easier. If it takes longer, then it starts fighting you every inch of the way.

With the shape in the boat being mainly in the ends, we would steam the forward ends of the plank and the aft ends of the plank – the planks could be 30 feet and we didn't have a 30-foot box to steam the whole length but 18 feet was sufficient to bring us back into virtually a straight area. So we would get the front ends of the plank, or the back end, if we were working on the rear of the boat, entered into their place first. We would clamp everything at this stage – we wouldn't drill and nail it or pre-drill it even, because drilling and countersinking

Figure 30: Fixing a steamed bow plank in position.

the nail heads weakens the timber and you could actually crimp or break the plank across the nail holes – so the idea was to steam it, bend it round and then clamp it. Then when it's gone cold then we'd drill and nail it. But having had a dry run first, everybody's a good idea where they're meant to be and what they're supposed to be doing. So when we actually say 'go' – though it looks like absolute chaos, and often is absolute chaos – and the air often blue, when somebody slips or has forgotten what they're meant to be doing or just through the sheer heat of the moment – it goes round relatively easily. (Figure 30) We'd got some basic clamps we bought from general hardware shops. We'd also got some proper shipyard clamps which we borrowed from Malcolm Brain's – he used to have a boatyard on the canal and so he lent us a great selection of clamps – and these are real non standard clamps and we referred to them as mongrel clamps. If we got halfway through something and the clamps we had weren't powerful enough or man enough for the job, then out would come these mongrel clamps. Some of these weighed 100 lbs each. Because they were such heavy and awkward things they took that much longer to set up and we didn't really want to use them unless we were forced to. But they had so much sheer 'grunt' behind them that they were really good for awkward jobs.

Finishing the planking

So the planking was quite a long and arduous process. The sequence of planking was not a straight forward one of laying them side by side. We left some gaps to be filled after, by what are called shutting planks. These were sort of wedging planks. If you remember back, I said we planked the sheerstrake and plus one. We also had planked the turn of the bilge, I think 3 planks or 4 planks were put in there. So we needed to fill the void between the keel and the bilge planks. Having planked out so far, the last plank to go is the plank we call a shutting plank. With all the best will in the world when you're wrapping planks around, you've steamed them and they're wet and then they dry out, there's shrinking or a small inaccuracy, so there might be a slight gap between one plank and its neighbour. The idea of this last plank being a shutting plank is, this plank you make – depending on how confident you are or how big are the gaps each side of the planks you are fitting against – you make it perhaps up to ¼ inch oversize. The idea is that when you start to sledge hammer this in place, it actually shoulders the other planks sideways. They're already nailed of course so you're fighting the nails as well. You'll shift the immediate plank each side sideways a bit and you might shift the one beyond that a little bit. We're really talking 1/16 inch a side – hopefully. If you're trying to shift more than that, then you should have taken the plank off and refitted it perhaps. The skill of the game is judging to make it a tight fit, but not so tight you that you can't get it in there. I remember with one plank, I was a bit generous in leaving timber on and we had one hell of a fight getting it in – we eventually got it in but it took us best part of a day with great worries that it just wasn't going to go. We'd try and get these planks wrapped round without steaming. You haven't really got time to steam them, but if they have got to be steamed, we'd steam them and wrap them around the plank adjacent and clamp them in place

Figure 31: 'Fairing in' the planking.

and let them get cold so that they'd already got the shape approximate to that which they had to take up. When they're cold and set then we'd take them off and start sledging them into place. Altogether there were 12 shutting planks in four strakes – you have one strake each side between the keel and the bilge keels and you've got another strake each side between bilge keels and the sheerstrake. Each strake was made up of three planks.

Finally we 'faired' in all the plank edges to give a good smooth shape. For this we used electric planes – even so, that was quite arduous too. (Figures 31, 32 and 33)

This completed the planking but one thing I haven't mentioned is fairing in the frames. Before any of the planking commenced, when we dropped off the old planking after fitting in the new frames, the frames needed fairing in, that is smoothing them up to a common even surface to take the planks. Because the old planking was so wasted and frayed, when we fitted the new frames inside the old planking they could only be plus or minus ¼ inch, so having got the old planking off now we had to fair in the all the new frames. This took a long time. I think we went a bit overboard in trying to get them perfect. We used straight edges and battens where we'd got curve – we'd wrap a batten round and where it came hard we'd fair it in. Around the top sides it's quite easy to work because you've got a vertical face – down underneath the boat was a real bitch – you're working a vast flat area overhead and we were glad to get out of that when we'd finished.

Another important thing I haven't mentioned – the planking originally was bolted and nailed. Every other fastening was a bolt – so each frame carried one nail and one bolt and this was staggered from one frame to the next. If you had a bolt in the upper edge of the plank on one frame it was on the lower edge of the next frame. The idea was

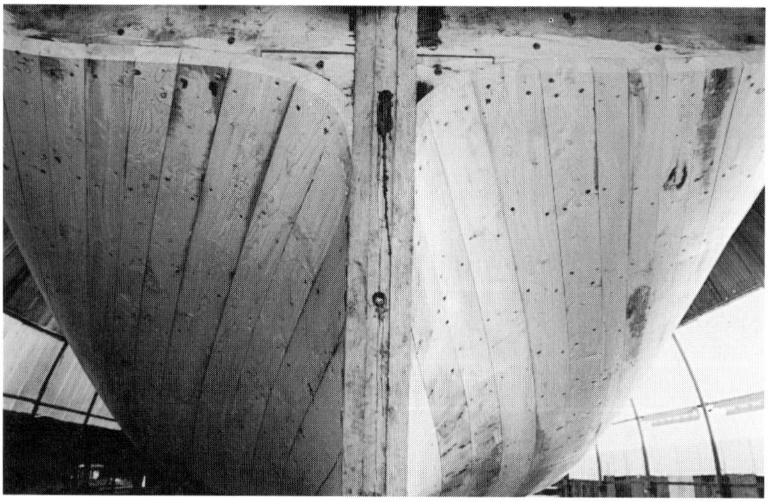

Above: Figure 32: *Spry*'s external planking complete and 'faired in'.

Left: Figure 33: Under-stern view of completed planking.

that with the boat working on and off beaches with heavy cargoes, and the beaches often quite steep, the boat would take a lot of working. A bolt is a positive fastener where a nail is just a friction fastener. Now if you've got a boat which is twisting and racking, a friction fastener can let go where a positive through fastening will not. Originally they were driven through and riveted on the inside face of the frames and then the ceiling put over the top of them. This makes sense when you have a boat in a working environment where the ceiling has to be sacrificial due to working stone cargos in and out. The ceiling's only going to last perhaps about ten years, so you'd need to rip it out and put another ceiling in. You wouldn't want to disturb all your through fastenings so the bolts actually terminated on the inside face of the frames. We didn't want to do that in rebuilding the *Spry* because the ceiling will probably last as long as the planking – if not longer – because we won't be working cargoes in and out. Therefore the replacement of outer planks in the future is of greater importance. She won't be carrying great cargoes on and off beaches so she won't have the racking and distortion. So we decided that we'd nail all the planks except the butt ends. The reason being here that if she were ever to be in a collision, the shock or jar could be enough to 'start' a plank. If a plank is going to spring off, it will start coming off from a butt or hood end. So all the ends we actually bolted through – not to the inside face of the frames, but to the inside of the actual ceiling. So if a plank needed replacing in the future we had access to all the bolt heads.

More on the internal planking

I've just mentioned the ceiling planking again. I described earlier on, how the vessel was planked inside all over as well as outside, but in telling how we tackled the making and fitting of the new planks I've mostly been talking about the outside planking of the hull. I'd like now to say more about the inside planking. The boat has an outside planking all around as a waterproof skin, inside all the frames there's another layer of planking and this is actually divided into two parts – there's the ceiling and the lining. The ceiling is across the bottom of the boat from the keelson to the bilge stringers and the lining is up the sides of the boat, above the bilge stringers. How the name ceiling comes about I'm not quite sure. It is all 2-inch thick planking again as is the planking on the outside of the boat. One would have thought it would be just as easy to put in as the planking on the outside of the boat was. We'd take our templates, in the same procedure. We'd get a spile, mark the size of the old plank off with a little chock of wood, onto the spile, transfer it all back onto a new plank and cut it out, with all the bevels so that the planks fitted edge to edge – the only thing we omitted was a caulking groove. Like in a barrel again, we had to bevel the edges of the planks for the full length so they'd fit against their neighbours. The inside skin of the planking isn't in any sense waterproof, it's only to keep the cargo from out of the frames and out of the bilge but it also gives us about a third of the strength of the vessel, for you're boxing the frames. Your frames are transverse and then you've got an outside fore and aft skin and an inside fore and aft skin.

We steamed most of these planks to help get them in position. It was much, much more difficult than planking the outside of the boat. On the outside of the boat you could get one end clamped in place and with the loose end acting as a lever, you had some 20 or 30 feet of leverage to bend the rest of the plank around. But inside the boat, the front end of a plank would be touching and the back end touching – but where the plank actually had to end up in between is perhaps 6 feet away from where it was lying in a straight line. The only way we could get at that, was actually to get two or three heavy blokes to jump on it. It sounds very crude but we tried pulling them out with ropes or jacks, we tried shoring them out with wedges, but the best thing was to get two or three chaps stood on them and start bouncing on them. This was near the floor of the boat. For the vertical sides, then we'd start holding onto a plank and getting our knees against it, gradually haul it in. Each bit we made, we'd try and get a clamp on it to give the blokes a rest and another grip. After all that messing around of getting the plank actually in place, then we could start thinking about nailing it and that was pretty straight forward as was the outside planking. The pattern of planking inside is a bit different from the outside of the boat. Against the keelson we left a 4-inch gap. This allows access down into the bilge to clear the limber holes. We've talked about limber holes earlier, they're to drain water from for'd, aft to the bilge pump, and of course if you've got dross or dust or mud or stones, coal, or whatever in the bilge, eventually these holes start getting clogged up, so you have to have access. This gap is closed with a loose plank, it was called a limber plank or a limber strake, which would be dropped in and spent most of its life in place keeping the cargoes from getting in the bilges as they were worked in and out of the hold.

So then the planking would start 4 inches away from the keelson and work out to each bilge stringer. Our actual method of working was to fit planks on the inside and the outside alternately, the idea being that we could reach through between the frames and put a clamp onto the outside of the boat to haul the ceiling in. We could also plank the outside of the boat reaching through with clamps to pull the outside planking in. So the ceiling, lining and the outside planking grew at the same pace, port and starboard. We had some problems in that the inside planking doesn't follow the same run as the outside planking, so in some areas the inside ceiling was way ahead of the outside planking. Luckily that was only in the ends of the boat and the ceiling planks would be touching there. But it really was a brutal job getting the ceiling and lining in.

Having got the ceiling and lining in, and the outside planking on, we could then rivet or bolt the butt ends of the planks right through and peen them over with washers inside the boat. And that really completed the basic hull.

Figure 34: Ceremony to mark the completion of the basic hull on 3rd November 1988. Brigadier Goulburn hammers in the final nail, under the supervision of Alan Williams.

The completion of the hull was a very significant achievement and the USNT arranged a formal Ceremony where Brigadier Goulburn drove in the last nail into the last plank fitted onto the outside of the hull. This took place on the 3rd of November 1988. (Figure 34)

The decks

The next step forward was getting onto the decks. We'd only got the small fore-deck and small aft-deck and in between the two you have the large open hold. Now the hold stands up above the foredeck and the aft deck by nigh on 12 inches. The coaming for'd and the coaming aft on the hatch, sits upon the deck so the procedure was we'd get the decking down and the cover boards fore and aft, before we could even start thinking about building up the sides and the fore and aft ends of the hatch. The first job then was to put a cover board around the boat. The cover board, as the name implies, covers the frame heads and it overlaps the outside of the boat by ½ inch, outside the sheerstrake and it also overlaps the inside of the boat. Now on the inside of the boat you've got your beam shelf and on top of the beam shelf you've got your deck beams. Now there's a void then between the beam shelf and the top of the deck beam in between the frames, and between the deck beams. The purpose for one, is ventilation in the cabins fore and aft, this is for ventilating the hull, not ventilation for the skipper's comfort. In the way of the hold, again you'd have loose planks which you could take in or out, one for cleaning down the bilge, two for doing inspection of the boat, and for ventilation if the boat's laid up at all. The cover board on the inside edge, doesn't actually line with the deck beam or plank at all, nor the lining or beam shelf. All it does is it actually spikes on top of each deck beam. So you've got a cover board of about 12 inches wide and the thickness of deck plank is 3 inches so you've got your cover board also at 3 inches. The outside edge has a bullnose on it, a rounding off of the edges, the inside face is left square.

The cover board, though we only have a deck forward and aft, actually runs the full length of the boat, but only as a dummy line. If you'd finished the cover board where the foredeck and the aft deck finishes, you'd have a butt where you've got two different modes of construction. So the cover board is an eye line – it's what your eyes are attracted to and emphasises the sheer. It actually runs right along beside the hatch but it's only 3 inches thick and 4 inches wide and it adjoins the outsides of the frame faces and is spiked onto these. That's the cover board. It's jointed just with butt joints and no more than that. Actually they butt onto the stanchions. The stanchions we put in after.

Going back to cover boards in the deck areas – these run round the perimeter of the deck. They are pierced by 4-inch square holes through which the stanchions, which eventually hold the bulwark planking, and the capping rail, are fastened. We butted the cover boards on the outside edge on the front end of the stanchions and on the inside edge on the aft of the stanchions. So you had a 'Z' shaped butt with a square inch hole pierced in the middle. The aim is to spread the joint and to be able to nail the cover board into a substantial piece.

Having done that, then we could get straight into planking the deck. The deck planking is about 5⅞ or 6 inches wide by 3 inches thick – with a caulking groove already machined in it. The under side of each deck plank has a ½-inch drip bead machined into it, the idea being, if you get any deck leaks or if you get condensation, the water doesn't run out and then drip off down the outside edge of the boat, it would actually drip off adjacent to where it forms. So, one, you can easily trace leaks, and two, you don't get fresh water running down into hidden areas of the boat and causing it to rot. It's a fiddle to do, I suppose we could have got a router cutter to do it, you can get them made, but we had a couple of old wooden moulding planes and used those. We spent a week and a half, two of us, just on the bench, running up beads on the bottoms of the planks.

The starting point for planking, of course, is midships, so we set up the first plank or rather the line, right through the middle. This centre plank on the boat is usually called the king plank and in many cases is a thicker plank, but that's generally on larger vessels. On the smaller boats and work boats it's not such a major piece. In *Spry*'s case, it's just a normal plank. It doesn't run the full length of the deck, foredeck or the aft deck, because the deck is pierced by – for'd, the samson post and also the crew accommodation hatch and – in the aft deck – you've got the skipper's skylight. So we just ran a straight edge through to give ourselves a line to work to. Because of this, the for'd king plank is made up of three pieces, and the aft king plank of two pieces. It's not until we get out to the width of the hatch and the skylight aft and the crew cabin companion way for'd, that we actually start getting full length planks. As we progressed outwards we also came then to the anchor winch legs which fit down through the deck, so whilst we were doing it, it was much easier to cut the holes out on the bench than actually cut them out in situ.

As we progressed then, we made all the apertures for the fixtures and fittings as we went, finally finishing off right at the outside edge of the cover board. The outer-most plank was only about 4 feet long – on the foredeck particularly, because the boat's got so much shape in it. The boat doesn't have any, what are called, margins. We've talked about the cover board. On a lot of boats inside the cover board there's another timber called the margin which runs around parallel to the cover board, into which the deck planks are often joggled, so as not to have feather ends on your deck planks. It's a more refined form of decking, unfortunately *Spry* is not that refined a lady – and also it's a lot more work. You can imagine a working boat was only built to work and not to be beautiful. It's only a natural thing to do. That really completed the decking. We trimmed off the aft end of the decking and put a bullnose on it where it went into the cargo hatch. The same on the aft deck where it came for'd into the cargo hatch. Also on the aft end of the aft deck we had to put a bullnose on just to finish it on the transom. This bullnose on the aft end of the aft deck is sometimes called the 'huddis'.

We didn't want to caulk the complete deck at this stage, we really wanted to leave that as long as possible so the deck could settle down and so we only caulked where the fittings were going to cover the gaps, and under the hatches. Where the hatch comings were going to go we would caulk the 12 inches of the end of the plank which would be covered up in the near future and coated them up with pitch, then cleaned them off.

The bulwark stanchions

That's the decking. We didn't go on to the main hatch at this stage – we then went on to putting the bulwark stanchions in and then on to planking up the bulwarks and the capping on the top. The stanchions are 4 inches square nominally, at the deck and they go to 4 × 3 inches up in the gun'l. The only reason that they taper is to give a bit of eye line, if you put something in which is square and parallel, it looks clumsy. Even if you just put a bit of chamfer on – it doesn't take much effort – it makes a world of difference to the actual aesthetics of the job. The inside faces of the stanchions have got a slight radius on them, on the corners, the outside corners are left square as they're covered with the bulwark planking. On each side of the foredeck there are 5 stanchions and there are 4 a side in the after deck. The for'd ones are quite difficult to put up because you have to decide how much flare to give. If you follow the line of the planking with your bulwarks, especially for'd, your bulwarks will end up sticking out further than the boat, the hull. Therefore, if you come alongside and come into collision, you can do that much more damage to your bulwarks. So you try and stand them up vertically, or sometimes even cant them inwards slightly – a bit of 'tumble home' we call it – so they don't come into contact with a dock wall, or other boats which may be rafted up alongside.

We had the knighthead for'ds – you remember the stem came up about 2 feet above the deck, and the knighthead and the hawse timbers also came up 2 or 3 feet above the deck so we had an established angle on the stem. At the hatch we knew we wanted it to be vertical or with a slight cant inwards, so we just had to establish a batten around a fair curve, port and starboard, and make up stanchions to fit. These were just fastened into the hull with trunnels, between the outer planking and lining, extending some 3 feet down into the hull. These were the only trunnel fastenings in the boat. The reason for this, I'm sure, is that with working the boat on and off quays and wharfs daily, when you're coming alongside, or leaving, taking a tow or tug, there's a good chance of damaging the bulwarks through collision, or through clumsy mooring line work. If you do break a stanchion, and if you've fastened it with great iron nails, you can't get the nail out without destroying the planking. Hacksaw blades I'm sure didn't exist, or certainly not in the quality they exist today, and anyway you wouldn't be able to poke those 3 feet down into the hull to cut through a fastener. So if you have a wooden nail or peg – which is the trunnel – driven in from outside, through the planking, through the stanchion, and right through the ceiling as well or the beam shelf, you can get on the inside of that and just drive it out. Or failing that, you can drill it out from the outside, using a wood auger and then draw out the broken stump of the stanchion. To do it any other way would be incredibly labour intensive.

Trunnel fastenings are a very ancient way of fastening boats together. It's perhaps funny that it's survived only in the stanchions. In *Spry*, in actual fact, the original butts of her stanchions, from when she was first

Figure 35: Alan fixing the for'd stanchion capping rail.

built, were still in her. When they raised her decks they cut the stanchions off to the new deck level and treated them as frames, then put in completely new stanchions alongside. So the butts of the original stanchions were still in place, still with their original trunnels in. None of them had been disturbed or replaced at any stage. You could tell this for the holes weren't messed up, they hadn't been messed around with. So in the years from 1894 through to when she had her decks raised she actually never broke her bulwark stanchions. It's interesting that they had made provision for replacing them though they didn't actually have to get round to it.

On top of the stanchions comes the capping rail, but before fitting this, we needed to strengthen the stanchions under where the fairleads are fitted on top of the rail capping on the transom and on the bows. So the fairleads aren't just bolted onto the capping, which is relatively loosely fitted on top of the stanchions. We transmit the load down onto the stanchions, and to do this there is a chock which is in effect dovetailed between the stanchions. This is the same both on the fore and aft decks, there's a chock like this, port and starboard underneath the fairleads. The forward ones have got a very convoluted type joint – it's dovetailed in two directions and when the rail capping goes on, it's locked in there absolutely solid, there's no way that it will move.

Having done this we could then put the capping rail on. (Figure 35) This is approximately 8 inches wide by 3 inches thick, with a large soft radius across the top, and harder radii on the sides. This is scarfed, being made up of a number of lengths, they're not full length pieces. The port side rail goes forward, right from the stem back to the end of the bulwarks. On the starboard side there's a cut out for the bowsprit. This cut out goes down into the knighthead and hawse timber by about 10 or 11 inches. This is where the bowsprit runs in and out, so the actual capping finishes just on the edge of the aperture and runs back again to the cargo hatch. The capping rail is fitted onto the tops of tennons on the tops of the stanchions which enter mortises in the underside of the rail. To hold the capping down – if you nail straight down into the top of the stanchions, you've firstly got a great risk of splitting the top of the stanchions. Another problem is that fresh water goes down the nail hole and into the end grain at the tops of the stanchions and rots the tops out. A third thing is that nails don't hold very well into end grain. We actually found an old handrail nail when we were cleaning the bilges out. What it consists of is a parallel shank nail with quite a large head. The lower end of the shank has been flattened out and is pierced by two holes. Now the idea is, you drill a hole in the capping rail, alongside the stanchion, and countersink it, you put the nail through the hole and then you nail this onto the stanchion with smaller nails transversely into the sides of the stanchion. This avoids the problems mentioned earlier and you are actually making a sheer fastening, instead of just a tension fastening.

The heads are sunk down into the rail and capped with a wooden plugs so you don't see them, although you can actually see them now because the plug hasn't shrunk as much as the rail has, so they stand up a fraction. We'll smooth them off again later. We made all the capping rail nails up in the forge. (Figure 36) It's one of the beauties of working in a Museum, you can do things as they were done originally. I don't know anybody today who's been repairing boats and has bothered to go to these lengths, because the easiest method and the quickest method is just to nail on top of the stanchions, or you could put a right angled bracket in, but you seldom see that either, because it needs two fastenings instead of one. It's one of the little quirks of working in a Museum.

Figure 36: Forging rivets.

We then fitted the cavels – cavel sounds very Spanish but I'm not sure where the name comes from. They're actually like large cleats on the inside face of the bulwark stanchions and they go between two stanchions and on which you can tie off ropes. So these are made of approximately 4 × 4-inch timber shaped to fit the curve of the boat and extend beyond each stanchion by about 8 or 9 inches over which you can hook or loop rope and tie things off. These are riveted on to the stanchions. These went on before the bulwark planking could be applied, because the rivets come through from the outside of the boat through the bulwark stanchions and then are clenched over washers on the inside face. The bulwark planking goes over the recessed heads of the rivets in the stanchions. Up for'ds you've got one cavel port and one cavel starboard – riveted through as I've said. But on the aft deck you've got two non-standard cavels – one port and one starboard. These aft cavels, which brace across the transom, are like great big corner brackets so that they extend 3 feet from the corner of the boat towards midships, to the transom stanchions and around to the side of the boat both port and starboard over two stanchions, and they finish 9 inches ahead of the second stanchion from aft. So these are quite major bracing pieces in the corners. It's really to take any loading from the fairleads on top of the transom from which you moor the boat up or tow her if she is on the beach and you are kedging her off the beach. You may well have an anchor a couple of hundred yards out astern, with the rope through the fairleads, up onto the winch forward, so you can winch yourself off backwards. The large size is to spread the load. So with these fitted, then we could tackle the bulwark planking.

The samson post and winch legs

The next thing we tackled was the samson post. This is the main structural point in the front of the boat, and it is from this that the boat would be towed when it needed towing. Also the anchor winch has its pawl bits on the samson post on the aft side, so if you'd got load coming on your anchor winch, a lot of the load goes onto the samson post. This is a solid lump of oak, 10 × 10½ inches thick at the top and nearly 9 feet long, because it actually goes through the deck right down on top of the keelson. That's nearly 5 feet below deck and another 4 foot 6 inches above. Down below deck it tapers away because you don't need the bulk down there. It tapers down to approximately 6 × 6 inches at the keelson and it's tenoned into the keelson. The bottom end of the samson post has an iron band fitted around it to prevent it splitting. If a load comes on the top end, there is a possibility it could split out at the bottom, but you must be talking about an incredible amount of load so hopefully it won't ever come into play. The iron band is the original bit of iron which we found with the boat when we took the old samson post out and so we put it back on again. We actually heat shrunk it on, just to get it as tight as possible. The very top of the samson post has its corners scalloped slightly to create something of a neck, so that if you've got a rope round it, the tendency is not to slide up over the top unless you give it lots of loose and this is for tying off on.

Figure 37: Bow, with samson and winch posts fitted.

The other thing about the samson post, it has a 2-inch square hole cut in it, only 1 inch deep, on the starboard side, in line with the bowsprit. Now this is to take a chock of wood – it's called a 'coke' – that's a short grain chock of wood that is inserted in the actual bowsprit and the idea is that it transmits load and helps to stiffen the mounting of the bowsprit. The bowsprit comes through the bulwark's and alongside the samson post and it is retractable. The idea being that when you come into harbour you would run the bowsprit back inboard, because one – it would save harbour fees because you get charged to your length in many ports and two – it's easy to do a lot of damage to things on the shore and to the boat in coming alongside with your bowsprit out. The bowsprit is then just bolted with an easily removable bolt at the side of the samson post, but that can't take much load and there's a lot of load coming onto the end of the bowsprit under sail, so to take the load you have this square hole – into which the coke fits which is in turn fitting inside the bowsprit which locates and gives a key, something like a keyway in engineering.

Coming to the anchor winch legs – they're of 9 × 4-inch section. Again they go right down to the very bottom of the boat. They're also tenoned at the bottom, this time into the ceiling of the bow and again they stand up some 4 feet above the deck. Above the deck they're parallel – below the deck they taper in their fore and aft section but they stay parallel in their thwart ship section. Again you don't need as much volume of wood down below but you need to have something to hold into the ceiling. Eventually they have bearings let into their after faces, but this comes later when we describe fitting the winch in place. Here we're just getting them in situ.

The cargo winch legs were next to go in exactly the same as the anchor winch legs, down onto the ceiling, tapering in the fore and aft section, parallel from side to side. That's the foredeck completed. (Figure 37)

Aft you have two large mooring bits, like miniature samson posts and there's one port and one starboard. And they're only about 4 feet long and are 9 inches square above deck. Below deck they taper down to about 4–5 inches square and are tenoned again into the ceiling and also with their top corners scalloped to prevent ropes slipping off. The only thing we did then was we put the groundings or the foundations for the hatches in place, the skipper's companionway hatch, his skylight and the crew hatch forward. We didn't continue with any further build-ups of the hatches at this stage, leaving this until we'd finished fitting out the cabins, as it would have restricted access.

The bulwark planking

Having got the rails fastened down and the cavels in, we could then plank up the bulwarks. But actually before we could do this, the chock at the mouth of the hawse pipe which takes the main anchor chain outboard through the bulwarks, had to be fitted. The hawse pipes are of cast iron and there are two of them. The original ones were still in *Spry* when she was recovered so they were refitted. The port one is a much more substantial one than the starboard because your port anchor is your main anchor, and you have your main anchor chain through there. Your starboard hawse pipe is only for your kedge anchor which is operated on a rope and not on a chain. That is only used as a casual anchor or for getting off lee shores or for moving the vessel up rivers. So then, you've got a much more substantial one to port and a light one to starboard. On the outside of the hawse timbers and knighthead there is this 2-inch thick chock, shaped into the stem and on the after edge it's got a bit of a sweeping curve. This is to add meat to the framing of the hawse timbers to take all this load. The bulwark planking doesn't actually go underneath the chock but butts up against it so that's why the chock was fitted first.

The lowermost bulwark plank is thicker by 1½ inches to the lower plank and 1 inch to the upper plank, the idea being that the top edge of the lower plank has a drip bead cut in again, not this time for any reason other than for looks, to soften the edge between the thicker plank and the thinner plank. Halfway down the plank you have a cavetto line carved in, which is a concave moulding, about an inch wide and

¼ inch deep running the full length of the boat. This is carved into the lower plank. This is also a thicker plank because it gets more abuse. If you imagine you've got stuff rolling round the deck, or you've got an anchor loose on deck and slides against it, a thin plank could be punched out, so we have a thicker plank and then 3 thinner planks above that, an inch thick, wrapped round and nailed in place with 2-inch galvanised handrail nails – and that completes the bulwark planking.

With the bulwark planking done, we could start putting some major pieces in. Things like the samson post, anchor winch legs and cargo winch legs.

The bulkheads

Having got the decking done and the bulwarks, we next put the bulkheads in at this stage. We didn't have any details of the original bulkheads so I made them out of approximately 6 inches wide, 1½ inches thick larch planking with a loose tongue in. These were nailed vertically. We fitted a grounding, a frame, to the ceiling of the boat and all around the perimeter, shaped the bevel with its – the face that had to be nailed against – plumb. The after bulkhead is nailed onto the for'd edge of the deck beam, but the for'd bulkhead is actually one beam in from the end of the deck so you have perhaps 3 feet of covered over cargo hold forward. We used 2½-inch galvanised nails. These main bulkheads form the division between the hold and the master's and crew's accommodation. We also put glazed windows in these bulkheads. They were not originally there of course, but it's the only way to let the public see and appreciate how the cabins are fitted out without them actually going inside them, which would not be practical because of the very restricted space. This is practically the only deviation from the original that we've actually done.

For'd there is also another bulkhead. You have what would today be called the collision bulkhead right forward of the crew's cabin, just forward of the anchor winch legs. The idea of this bulkhead nowadays, is that it is a water tight structure, so that if you are in a collision, if you've holed the bow, water won't come rushing into the front of the boat. In *Spry*'s days it would probably have just been called the forepeak bulkhead and it was used for the storage of bits and pieces and had a small hatch through for access.

The transom name board

After finishing the stanchions and capping rails and their fittings, we next made a new name board for the transom. The original one came with the boat when we rescued it from Worcester. (Figure 38) It had obviously been altered in the past, when they had raised the hatch sides.

Above: Figure 39: Naming ceremony on 25th May 1990. David Lawrance, Chairman of USNT, unveils the gilded transom name board.

Right: Figure 38: The old name board.

They had reduced its length to accommodate the tumble home which they built into the new bulwarks. Also, she'd had her port of registry changed from Gloucester to Bristol. So we decided to carve a new name board altogether, using the original one as a pattern. Luckily, underneath the layers and layers of paint and tar, you could still see a vague outline of the original port of registration, carved in. What they did when they came to re-register in Bristol was they just hacked off the face of the plank with an adze and over-carved 'Bristol'. They managed to keep

Figure 40: Mrs. Pat Ellis and Mrs. Jean Knipe, daughters of the last Master of *Spry*, at the naming ceremony, by the unmounted rudder.

two letters of the original – the S and the T of Gloucester – and then superimposed B, R, I and O, L on the ends.

Having found the outline of the original letters, I could then take a tracing – it was very interesting to see the original spelling of 'Gloucester' which was 'Gloster'. Also, on closer inspection, we could make out the little motifs on either end of the name and on the port of registration and when I was cleaning those out to reveal their full depth and form of the carving, I came across what must have been some of the original flakes of gold leaf right at the lowermost layer. So when she was launched she had this 'SPRY, GLOSTER' all picked out in gold with little motifs on either end. It was great fun to be able to carve the figures new, afresh, and then get a signwriter to actually gold leaf them and this we did. It was a very proud moment when we had a naming Ceremony later, on the 25th May 1990, when we unveiled the new nameboard, gleaming in gold and black, just as she must have had originally. (Figures 39 and 40) Lots of comments were passed on the spelling of 'Gloster', luckily I still had the evidence before me, having kept the old name board.

The main hatch coamings

After this we went on to make and fit the main hatch comings. The frames in the way of the hatch stand up some 9 inches above the cover board. This was a quite difficult job. The hatch sides consist of 2-inch thick planks on the outside and 3-inch thick planks on the inside and the capping 10 × 6 inches. It's quite a substantial lump of timber. This 10 × 6-inch capping has a 3 × 2-inch rebate on the lower inboard face. It also has a 2 × 2-inch rebate on the lower outboard face and a 2 × 2-inch rebate on the top outboard corner. Inside the boat we have the ceiling coming up to the bilge stringers and the lining coming up to the beam shelf – then we have the 6-inch gap before we plank up to the top of the cargo hatch, which took two planks. The topmost plank stands 2 inches above the top of the frames. We then had to pull round the capping – there's quite a bit of bow in it. We didn't have a piece of timber with sufficient section to cut the curve in it so we had to edge bend this 6 × 10-inch section. We used the face of the inboard plank as a former to pull it against. We did this over a couple of days to give the timber time to adjust to the shape. Then having got the capping in place, we could wrap the outside plank around and located it in the lower outboard rebate of the capping. The inboard plank had located already in the inboard lower rebate. Having got it all in place, then we could fasten the whole caboodle together. So your outside plank was then nailed in horizontally into the rebate and the inboard plank also nailed in horizontally. (Figure 41)

Figure 41: Paul Turner nailing the top outer plank into the main hatch capping.

The cargo hatch beams

Next we came on to the cargo hatch beams. Originally there were three in the boat when she was built, but when we acquired her, there were only two. One of those was original but the other one was a lump of what seems to be greenheart, I'm not quite sure. The original one had its original tonnage mark and registration number carved upon it and that was in the position of the centre beam. Now originally, your tonnage beam is your main beam in the boat – therefore it should have been the tabernacle beam, which the mast pivots upon. On closer inspection of this beam with its marks carved upon it, we could in fact, see where the tabernacle was originally mounted and its adjacent supporting knees. Somehow from its original position it had been moved one position aft – to being midships in the hold. I can only assume this came about when *Spry* was having her decks raised or when they put the new keelson in. They would have had all the beams out of her and obviously they kept the original tonnage beam. Probably its ends were a little bit rotten so they trimmed the ends off and moved it to a narrower part of the boat.

I particularly wanted to use this old beam again but rather than put it back where it should be, at the tabernacle, and build the tabernacle back upon it, I felt it best to keep it in the position where it was found, which is not such a stressed job and I put a new beam in for the tabernacle. Having decided this we then repaired the rotten ends of the old beam and cleaned them up so that it'd fit close to the side of the boat. We then cut two new beams, one for the tabernacle and one for the after end of the hatch. These are about 14 × 10 inches in section with 2 inches of camber in the top edge, being straight underneath. These were lifted up into the boat, just Paul and myself, again with a sheer legs and chain block. The longest one was about 16 foot 3 inches and the shorter one was about 15 foot 2 inches long, so you can imagine they were quite meaty bits of timber to handle.

These are held in place by hanging knees and lodging knees. They don't sit upon a beam shelf in the conventional way of beams. The hanging knee is a big iron bracket which is fixed underneath the beam, bolted through the hull and through the beam. Lodging knees are horizontal brackets – one on the for'd side and one on the aft side. They brace the beam thwartships. When we got *Spry* there were some hanging knees and lodging knees still in situ, but as I said earlier, one of the hatch beams, the after one, was totally missing. There were the remains of half of one of the hanging knees on the port side. This had been cut off – obviously they couldn't get the bolts out so the easiest way to do it was to cut the thing off. On the starboard side there were no remains. In all, we had about 50–60% of the knees – the others we had to use salvaged ones we got from wrecks around the estuary. These had to be re-forged to fit the angles that the *Spry* set up. Also they had to have a set put in them where the hanging knee drops down over the beam shelf against the lining planking.

To do all this forging, I had to build a forge. The only forge in existence at the time was that up at the top of the Blists Hill Museum site in the blacksmiths shop, which had hand pumped bellows. To have somebody pumping air continuously whilst doing large bits of iron forging was really not on. Also I wanted to be able to offer things up in position to check the shapes, doing a job two or three hundred yards away from the actual boat was quite inconceivable, so I built a forge by the boat.

These were quite heavy gauge brackets — the section of the heaviest one was about 3 × 2 inches. To put a side set in a bit of 3 × 2-inch wrought iron and do it over a very short distance with quite a primitive forge, no steam hammer and very few tools – also with very little experience on my part, was quite a task, but we managed it anyhow.

All these then had to go away to be shot blasted to clean off the rust and scale. Then they were sent off again to be galvanised. Joseph Ash and Lacy, the galvanisers, offered to do the galvanising free as a contribution to the project. We've been fortunate and have taken full advantage of many such offers throughout the whole project. They have a plant in Telford with an enormous vat. They do dip-galvanising there of larger items, but the smaller stuff – nuts and bolts and nails – had to go to a more suitable plant for these, in Birmingham, where they have a centrifugal tumbler or 'cracker' to shake off all the excess zinc.

When I picked up the iron brackets from the shot blasters, I could see that even some of the original brackets were actually second hand – even 100 years ago there was second hand iron. This applied really only to the lodging knees, the hanging knees seemed to have been made from new for the job. You could see that the lodging knees were second hand iron which had been forged together. You could tell because where you had the two arms coming into the corner of the knee where they were forged together to give a lot more meat in the corner of the knee. In the back of the forging you could still see redundant holes from its previous life – what this had been 100 years ago, heaven only knows. It was nice to find that the tradition had repeated itself – I was going out salvaging iron knees whereas the boat builders 100 years ago had been doing the same thing.

They came back with a good heavy coat of zinc. The reason we decided to galvanise everything was because we couldn't make all the fittings, particularly nails and rivets or bolts, out of wrought iron, which is more corrosion resistant, but we didn't have the raw material. So to give them a bit more protection we decided to galvanise them. The original wrought iron ones we decided to galvanise as well, as it wouldn't do any harm and it could only do good. This is the theory behind this.

All these hanging knees and lodging knees are fastened in by ¾-inch bolts. Bolts in the traditional sense, as I've explained earlier, were just a bit of bar driven through and peened over a washer or peened over a piece of iron. I decided to actually fit bolts in the modern sense, put a bolt through with a nut and washer, for the reason that if we need to do repairs you can undo the nuts easily, and back the bolt out to replace a plank. Also bolts keep their pressure. The boat's rigidity depends upon there being no shackling or no play. With bolts, the conventional sort where they are just peened over, they are almost impossible to tighten up at a future date, but if you've got a thread then you can come along with a spanner and give the nut another turn – and this is what I did. I couldn't buy bolts to do the job exactly – I had to buy high tensile steel machine hold-down bolts. They have a square head with a square shoulder under the head. It was the square shoulder under the head I actually wanted. I didn't want the square head, so I just put them in the forge and panned out the head to give me a large domed head to put against the larch planking. In all, about 120 of these bolts had to be forged up, shotblasted, cut to length and have their threads extended to suit the job. All this took a remarkable amount of time, but it pays off in the long run. We have a secure job now. (Figure 42)

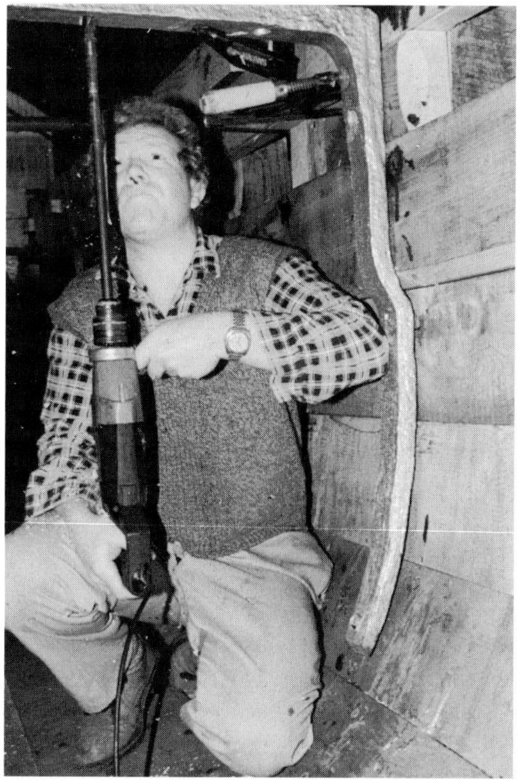

Figure 42: Alan Williams drilling into a hatch beam, through a galvanised hanging knee.

Another thing I did to add strength to the vessel, was to fit two stanchions under two hatch beams. These were not originally fitted and would never have been needed for in her working days she relied on the 100 tons of cargo to keep the bottom pushed out and straight, but with the boat now going to be forever light, there is no stiffness in the bottom. The bottom of the boat is just like a big flat raft. So if she ever does float or if she is ever chocked up on dry land the tendency is for her bottom to go upwards. To counter this, I put the stanchions between the hatch beams and the keelson. The main hatch beam – the one with the mast tabernacle – doesn't need one because its got the tabernacle fulfilling the same purpose.

Having got the hatch beams in, the boat was structurally 100% together now. There was nothing left to do structurally on the boat at all. Having got this far we then set our minds onto the fittings and then the spars.

The anchor winch

The next job we undertook was the anchor winch. The complete original winch was still in the boat when we got her from Worcester. All the woodwork was in a pretty shocking condition but the ironwork, luckily, was basically fine – though rusted and pitted and well scaled with rust. With some careful cleaning up, it was found to be O.K., though some of the teeth on the ruffle were quite blurred. The ruffle is the collar in the middle of the winch which takes the pawls to stop the winch from running backwards. If you think of a fifteenth-century courtier's dress, they'd have a great ruffle round their necks with vertical pleats, and this is where the term comes from, or where the likeness comes from anyhow.

The winch shafts were O.K. but found to be very worn in the actual bearings. We didn't want to put them back in, in this worn condition so we had them built up with weld on the bearings. Unfortunately, the Museum had no machine big enough to put the shafts in, especially as the large cog was still attached to the shaft. I was very loath to disturb it, it had been on there for 100 years. To think about disturbing the keyways or the shims that were locked in there, to try and replace them, to get them back exactly right, was more that I dared contemplate. So this shaft with a cog on it wouldn't go in any lathe. What I did, I lapped the bearings in by hand, taking the bulk off with a portable grinder, then with a file, working finer and finer. The bearings were very crude, they are only cup bearings and only half cup at that. The anchor winch only ever pulls in a for'd direction, therefore you only need a cup on the for'd facings. The aft face is an open gate which is closed off by a flat sheet of metal, which hooks onto . . . if you imagine a gate latch where you have a rising wedge and then a sharp drop off to a vertical face, and the locking latch bar drops down behind it . . . this flat sheet of metal drops down behind the faces just like that. The bearings were not lined, they were just steel – steel bearings – steel on steel. Nothing more clever than that. They would use lots of grease. Rusting wouldn't be a problem. If something's used on a daily basis – corrosion won't get in there. It's not a high performance bit of engineering, its revolutions would probably be only 20–30 a minute. That's flat out perhaps!

The drive cog – that's a small cog with a pinion – was quite worn too but we still had sufficient meshing for it to operate fine. Again the bearings were badly worn on the top shaft, these we built up again by grinding, filing and lapping them in. These also had cup bearings, but they were two sided, split bearings, again, only of wrought iron. The bearings themselves, we had to make new, the cup bearings. The original ones had worn down to be paper thin, which you could imagine, after driving away for so many years, just iron on iron.

Figure 43: Lifting the anchor winch drum aboard, using sheer legs and block and tackle.

As concerns the woodwork – we had to start from scratch. Luckily, as I said earlier, we had all the original woodwork even though it was in bad condition. The main barrel went together nice and easily and we could rivet that through from side to side – the main barrel is split in two halves and clamped each side of the square shaft with a rivet which runs right through the middle of the shaft, out to the other side and peened over.

The two warping drums on either end posed a bit more of a problem. They're wooden tapering barrels with a lip on the outer end, about 11–12 inches long and 10 inches diameter – made of solid oak. The lathe in the Museum workshop was, one, far too fast running and two, its bearings weren't capable of coping with that sort of stress – so in the end we went into the Museum Victorian machine shop and used one of the metal working lathes there – very cumbersome, because we couldn't run it up at the full speed, which wood likes, to take the shavings off. We could only run at the speed which these Victorian machines would work. Russel, who was a volunteer up at the carpenter's shop, did a wonderful job of turning these. We picked particularly snarly bits of oak, the main reason being – if we had a nice and clean run of oak, with no knots in it – its tendency would be to split. You can imagine, with a square hole chiselled right through the centre of the winching drum, that's your drive and you try turning the drive shaft in a square hole, it tends to split the logs, so we had particularly snarly, knotty pieces of oak which gave a more random grain with the interlockingness of the knots – though it made it a real sod to turn – it made it hold together very well. These had iron bands on their inboard and outboard ends, also to help against splitting.

That's really all the winch consists of. It's a horizontal barrel winch – you have your ruffle in the middle – your port side is the side on which the anchor operates with a chain. The starboard side is really redundant, the port side being the main anchoring side. To stop the anchor chain chafing away the wooden winch barrel, you have fashion pieces of wood about 2 inches thick, on each face of its octagonal shape. In the joints of these fashion pieces, there are large iron staples called 'thrapping irons', which again stop the corners getting worn away and give a riding surface for the chain.

To get all this lifted up in place – we manufactured it all on the site in the workshops – we rigged a couple of telephone poles again as sheer legs for a chain block. I don't know how heavy it was, but it was all two of us could do to lift it, so you must be talking 300–400 lbs of weight. It went in without any trouble at all and it worked beautifully. (Figure 43)

Actually, before we could insert the barrel into its position we fitted what's called the pawl box, which is a cast-iron shallow tray in which 5 pawls pivot. This is located on the aft side of the samson post. The original pawl box was in reasonable condition – two of the pawls were a bit rough but usable – the other three weren't. For these I had to make up patterns to get new cast-iron pawls cast and drilled out and fitted. 5 pawls perhaps seems a bit excessive but there's a lot of strain comes on the winch, particularly if you're anchoring in a fairway where there's a 4–5, maybe even a 6 knott current flowing past you. You don't want any mishaps so 5 pawls gives plenty of back up.

Having slid the barrel in place, and locked the bearings in place with the closing plates behind, to fair in the aft end of the winch legs, there are two cheeks of oak which again are just bolted in place. This allows rope to render around without fouling on the bearings and bits and pieces – to prevent snagging.

The cargo winch

The cargo winch is of a similar design but it has a two-speed drive and has its 'nigger' heads on the upper shaft, not on the lower shaft. The 'nigger' heads are the warp barrels at the end. We have no physical evidence of what the original cargo winch looked like, but we do have a photograph, taken in 1933, of *Spry* when she was a dumb barge with a winch on her foredeck, other than the anchor winch, which gave us the approximate specification, which was a top and lower barrel with a large cog and had a drive pinion which could be slid in and out of gear to give you two drive speeds. I say two drive speeds – you could take the lower barrel out of gear and then you could just operate on the top barrel. The lower barrel only had the one speed. The cogs we found around the Museum scrap pile. The larger one came from a winch which looks, from the teeth of it, never to have done any work at all – they are still sharp and crisp as the day they were cast. The pinion which went with it had half a dozen teeth missing. How this could come about we don't know, but we managed to find another pinion which meshed fairly well with the cog. We then got two bits of square 1 ¾-inch iron shaft, had bearings machined upon them and then built up the barrels on those.

With these we had just about completed all the major stuff on the bow of the boat.

We fitted the rudder next, but before getting onto that it will be interesting to tell about my scavenging trips.

Scavenging for parts

Over the past two or three years I'd heard of reports of wrecks up and down the River Seven and out into the Bristol Channel and at odd weekends I'd investigated these sites with the possible idea of salvaging parts which we could then later employ on *Spry*. I'd carry back one or two pieces each time, if I could actually lift them or they were free of any wreckage – put them in the car and bring them back to the Museum. I needed another, major, scavenging trip to pick up bigger items I had seen around the sites and I 'Shanghai'd' my father to give me a hand on this. We spent two days down at Lydney, Kempsey and Tewkesbury, salvaging bits and pieces. At Lydney we got 7 or 8 large iron knees – big iron brackets – which acted as lodging knees or hanging knees to support the cargo hatch beams. I got 3 rudder pintles, which were rudder blade pintles rather than stern post pintles. I also had a major find in getting a primary bilge pump. I found one piece of the pump on one side of the beach – the lower part of the pump I found 100 yards away up the beach. The two had become separated, obviously when somebody had wanted the barrel of the pump – the barrel originally being in bronze and more valuable. So they'd cut the bolts top and bottom, separated the two cast-iron bits and walked off with the bronze as very saleable scrap. These two cast-iron bits had probably been rolling about the beach for fifty years. I didn't know their form, or the style that this pump took, but I recognised the top and the bottom bit had such similarities in bolt positions and design to be off the same piece. These all went into Dad's trailer to be carted off up home.

The next wreck was at Tewkesbury. I'd heard of a barge in behind the lock there. On investigation it turned out to be a Thames Barge – a long, long way from home. She was called the *Monarch of Rochester*.

She was originally owned by the Flour Mills at Tewkesbury to carry grain from Avonmouth up to Tewkesbury. Whether she finished up motorised or as a dumb barge I'm not sure, but she wasn't sailing when she came over to this part of the country. Upon her, we found there were 5 bilge pumps, of a style which was typical to barges the country over, also 2 rudder pintles from the stern post, which I removed – they're now on *Spry*. We were able to salvage 3 of the bilge pumps, managing to get them out without too much difficulty, though the darned things weighed 150 lbs each. Staggering through the mud and through the undergrowth with them, to get back ashore, proved quite a burdensome job, but it was well worth it.

We then went to Kempsey, which is between Worcester and Upton-on-Severn. There there is the wreck of a barge called the *CFH* – she was a North Devon Barge. She became a houseboat after finishing her working life for the local boatyard. Then she was dumped, probably in the twenties or thirties, on the banks of the River Seven to stop erosion into the farmer's field. Actually, there were originally three of them in a row. Two of them have completely disappeared. This last one, the 'CFH', is missing so much now, particularly after we'd been there! On a trip earlier I had previously salvaged the anchor winch. This was a pump windlass type of quite a rare pattern, particularly these days and I had used this on a boat called the *Garland Stone* down at Plymouth. Whilst I was there on that occasion, I noticed she had 4 or 5 large hanging knees or lodging knees, which were ideal for *Spry*. So, with my father, we spent half a day with a hack saw cutting through the bolts and lifting out these iron knees – having previously of course, got permission for all of these things. Luckily, my father knew the farmer upon whose field this boat lies.

So we rescued all the bits and pieces and got the Museum lorry to do a pick up from three caches we had made of the salvaged wreckage from up and down the River Seven. We picked them all up in a day and carted them back to the Museum.

The rudder

Having got all these bits and pieces, we could then start thinking about the rudder. She had no rudder when we acquired the boat and photographs that we have of *Spry*, don't show the rudder at all. All that we could really say was that the rudder would have most likely been a similar shape and style to those on other trows we did have knowledge of. We just had to work by eye from there. The rudder post is of oak. The blade is of pitch pine. We had got some salvaged bits of pitch pine around the Museum. The pitch pine boards of the blade were tenoned into an oak shoe at the bottom of the rudder to take grounding and

Figure 44: Upper part of rudder after mounting.

Figure 45: Lower part of rudder, showing pintles and pin.

wear and tear. It was all riveted together with iron rivets, right the way through the blade, riveted on the inside face of the rudder post.

The aft edge of the stern post on the boat has a full width concave gouge or groove, running its full length and there is also a concave groove on the forward edge of the rudder post. This is to take, or to allow for, the bearings. There are three pintles on the rudder and three pintles on the stern post and a removable iron pin the full vertical height of the rudder. It then drops down locating all the pintles together. The object of this is, if she strikes a rock or when she grounds, the rudder can ride up rather than break or wrench the fittings off the boat. To do this, yet still have a close coupling of the stern post and rudder, one, the stern post bearing has to ride up and down the face of the rudder, and two, the rudder bearing has to slide up and down the face of the stern post without fouling. The pin and the pintles in the gouges allow this.

Figure 46: After deck with Alan holding the tiller, and showing the skylight from the master's cabin, companion hatch and bilge pump.

Having made the rudder, the pintles then needed quite a bit of re-fashioning to suit our *Spry* – the straightening out of bends – the altering of angles. Most of all, the bearings on these pieces of salvage were quite worn and of varying sizes as they came off various boats. So we had to bore them out, bush them, and then bore them out again so that we could end up with a uniform size, and a uniform clearance. When it all came together, the pin that the rudder rides upon – that also came from Lydney – actually slid down in place without any trouble at all. On the boat now, you can easily shift the rudder with your little finger in spite of the weight – there's so little friction. (Figures 44 and 45)

Having made the rudder and hung it, we then had to make the tiller. Again, we had no evidence of what this was actually like. All we knew was, that it had to be such and such a length to clear the two mooring posts on the aft deck, also to clear the skipper's cabin skylight, but then, not be too long as to foul the skipper's companionway hatch. So we had a length and a height and how we arrived at the in-between shape was just by eye. It is a very subtle shape – quite pleasing, the actual sweep of it. (Figure 46)

The mast and spars

We'd been out earlier – about 18 months earlier – with the Brigadier. He had two contacts where the Douglas fir for the mast might be obtained – one of them was the Powis Estate and the other one

Figure 47: Felling a Douglas fir for masts and spars.

unfortunately, I can't remember. We also had help from the Forestry Commission which gave us the Douglas fir for some of the spars. That came from the Wyre Forest just west of Kidderminster. So we went over there with the Brigadier and we selected 8 trees to do the job, (Figure 47) that's for the mast and top mast, bowsprit, stays'l boom, main boom, gaff and tops'l yard. The longest one was about 55 feet which is the mast. The boom is almost as long at about 52 feet. At this stage we didn't know the exact lengths we needed to make the spars. I based the whole sail plan upon a painting that was done at the turn of the century. It's the only evidence we have of *Spry* in her original form. It now resides in the Folk Museum in Gloucester. Its owner, a Mr. Joy, could not be traced, but the Curator of the Folk Museum very kindly agreed for us to photograph the painting. Having got photocopies of this painting, scaling it up on the drawing board, it's remarkable how it fits the boat, plus or minus 6 inches. Scaling from a 120 scale up to full size – 6 inches is not much of a problem. The detail in the painting was also extremely correct – the stanchions were in the right place, so were the anchor winch, the stove pipes, the skipper's skylight – everything fitted the evidence we found on *Spry*.

The other thing was, somebody had been over the painting after it was finished, with a pencil, and drawn in all the running rigging. It's a very personal thing is running rigging – every boat owner or skipper has his own quirks or idiosyncrasies in rigging a boat. The rigging was all lined in – running behind this spar – in front of that one – running over the top of this fitting or underneath that fitting there. So this suggests to me two things, firstly I think the painting was done from a photograph – it would be lovely to find the photograph it was done from. Secondly, the rigging was drawn in by one of the members of the crew, if not the skipper himself. The painting I believe, was commissioned by the skipper/owner of *Spry* and it seems reasonable to assume that it was actually him who put this later detail in. With all this evidence I felt quite confident in drawing a sail plan around this painting and this gave us our spar lengths.

For moving 55 feet lengths of timber by road, I thought we'd have needed at least a Police escort – but they were not that fussed about it. Apparently it's only when you go over 14 feet wide that you need a Police escort and I suppose also if you go over a certain amount of overhang which you are limited to on a lorry. But you do have to put these red and white triangles on the back of the lorry and things like that. As it was, we had to have a flashing orange light on top of the lorry – a flashing orange light on the after most part of the load, and a red and white triangle, both in the fore and aft line and athwartships so you could see it from the side as well as from the back. We then had to pick a route that didn't have too many tight corners for wiping out telegraph poles or telephone boxes. Luckily we managed to pick a back road and with a bit of care on some of the corners we got everything back here on a really foul day – it poured with rain.

We did this eighteen months at least – nearly 20 months – before we actually needed to cut them into spars. The idea being to give them some chance of drying out without cracking too much and to let them normalise or settle down before we started chopping into them. To help this, we de-barked them, then painted them all over with a wax slurry – to give them a coating to slow down the drying process. We also put them in under the North side of the shed so they wouldn't get any direct sunlight, they'd be kept cool, this is the dampest, coolest side that we had. They held up very well. In the Summer, occasionally you'd hear a crack coming from some of them – even today they've got cracks of up to ⅜ inch wide but it's nothing to worry about. If you could get the perfect bit of timber, and guarantee that it won't have any cracks or any knots, you could get away with a mast size of half the dimensions. Cracks are unavoidable so you just double the scantlings.

With regard to the spars – the way we were going to do them, rather than just adze them out right from the word go, was we'd set them up as level as possible, set up a centre line with a string line, draw out our width and our tapers to the spars and then run down that line with a chain saw, peeling off both sides of the tree, then turning it over 90 degrees, planing up the surface which we'd chain sawed and marking out again our widths and our tapers. Then I used a large circular saw which could cut 6 inches of depth – so we'd cut down the other face with the circular saw, turn it over 180 degrees, clear the other face off with the electric plane, and then cut down again with the circular saw – in fact it left about an inch of meat in the middle, so that last inch of meat had to be done with a rip saw. This gave us a square block and we then kept taking the corners off first to octagonal, then 16 sided and then rounded it all up. It took about two weeks to make each spar – that's just the raw spar without putting fittings or the finer details upon it. Actually, Paul Turner did all of this as I was working away from the Museum at this stage. I came back for a week in every three. We would then rough out a couple of spars with the chain saw and get stage one and two over and done with. Paul would then carry on, taking the corners off from square to eight sides – then sixteen sides – and then to round again, but leaving it square in the required positions. That work was done and completed by May 1990.

Unfortunately we started having hot Summers thereafter, and though we'd got them stored in the North side of the shed, the spars were beginning to work and dry out faster because we'd removed so much of the material. We coated them up with 3 or 4 coats of raw linseed oil and then gave them a coat of boiled linseed oil, the idea again being to stop them drying out or at least to limit the drying out.

Having done that, we left them in this state for about 12 months before we started putting the fittings on. I'd been thinking about the fittings and going over designing them, in my head, in the intervening 12 months. They consisted of iron bands with lugs on, bolts through the spars, the cross trees and trestle trees which are where the rigging hangs off from and the top mast sits upon. Also the cap band which the topmast telescopes up through, and the ruff around the bottom of the mast which the boom jaws sit upon. We had to make the boom jaws, the gaff jaws and a tumbler on the inside of the gaff jaw to take the loads at the varying angles that the gaff will actually sit at. (Figure 48)

Figure 48: Finished masts and spars.

Having made all these and painted them up, they then had to be leathered and for this we got a hide and a half of leather. One hide for the upholstery in the skipper's cabin which we will come on to later, and the half hide was for chafing pieces on *Spry*'s rig. The hides we got were nice bits of chromed leather, full size, unskived and fully packed with grease. When leather's cured, a lot of the grease and oils go out of it. It leaves the surface open for rot and degradation in the future, so for external use it needs re-stuffing, as the term is called. This is saturating the leather with grease or tallow, or silicon grease – one, to keep it supple and two, to exclude all moisture. So the jaws of the gaff and the boom were then lined with this leather.

Regarding other bits and pieces on the spars – there were iron bands on the ends of the bowsprit and spars, also sheaves through the ends of the bowsprit and spars. There are also bands on the stays'l boom, and a gooseneck on the for'd end of the stays'l boom. On the main boom we have an iron band which is the main sheet band, the main sheet comes from it – and also a band right at the very aft end of it. I don't know what it's called but it's really for tweaking the boom midships when you're trying to tack the boat in light airs. We forged all the bands here in the Museum. Some were old fittings from other barges which I re-shaped and re-fashioned. The main sheet band is one in particular. The rest we made here, we forged them up and put the eyes and lugs on them. We tried where possible to use wrought iron. The Museum, at that time, in the new Ironworks, was developing the technique of re-working wrought iron scrap into new finished products. They were not producing the sizes I needed but I would go through their pile of scrap awaiting re-working, selecting bits of wrought iron which I thought I could use and re-fashion. So we've got bits of old railway bridges and roof trusses etc, built into the *Spry* in varying forms.

The next job was fashioning and fitting the chain plates and dead eyes. The chain plates are the metal straps which bolt to the hull on the outside, abeam the mast. There are four either side of the vessel, each with a large lug on the top which is forged in. This was my first try at doing fire welding. Though they're not beautiful, they went together relatively well, better than I initially expected them to. Again, these were made up from wrought iron scrap. These were from bits from a railway bridge – actually a foot bridge. Coming on to the dead eyes, these are circular, 7½ inches diameter × 3 inches thick, lumps of elm, turned up in a lathe and they have three holes pierced through them. The holes then have scores, which is a recess running into the holes to render rope around and this the method for tightening up the rigging on each side of the boat. Again, there were 4 of those stationed on top of each of the chain plates. Around each of the dead eyes there is an iron dead eye band, this had to be forged up with a bolting eye on each end and to be made of such an exact size so that when they clamped up, one, they gripped the dead eye nice and tightly, and two, they actually came close to gripping the top of the chain plates. When I made them up at first, I found I had to go back through the lot of them and fine tune them, because though they were plus or minus ¼ inch, that was just too much leeway. In the end they all went on and nipped up beautifully tight. We soaked the dead eyes in linseed and 'rotproof' compound for 48 hours each. They're out in the elements all the time – they get little maintenance and moisture hangs around in them, around the rope, so this is just a preventative thing to keep the moisture out.

As I mentioned earlier, in the spars there are also sheaves, like a pulley wheel set into the actual spar. There's one at the end of the bowsprit too and another one in the top end of the top mast. These had to be galvanised and bronze bushes inserted and were fitted in slots cut through the spars, with their respective scores and with their bearing pins. Also in the bowsprit and the topmast, in the lower end of each, there has to be a hole drilled right through the butt with a score running into it. This is to facilitate heaving up the topmast, because the topmast could be raised and lowered by the crew and to do this you needed to render a rope through the butt end of it. The bowsprit the same. The bowsprit would have been run in when she was in harbour and out when she was at sea, as I said before, to save harbour dues and to avoid fouling other boats in the harbour. To secure the topmast when it is up in sailing position, there is another fitting, a 1¾-inch square iron bar, which is put through the mast – it's called a 'fid' – the fid then rests on the trestle trees. To raise the topmast there are rattlings on one or sometimes both sides of the mast, to climb up to secure or release the fid. The rope used for hauling it up is unrigged and sent below again – but as the bowsprit would have been in and out quite regularly the rope would have been left attached. All the leads and the angles of these holes and scores had to be just right.

We also needed to fit the lugs in the bulwarks for'd to which you attach blocks and tie off the ends of the ropes for all the rigging. There also are the whiskers for bracing the bowsprit, they had eye bolts on the outside of the bulwarks. The rope for hauling out the bowsprit needed two eye bolts, positioned in the right place, so that when you put a block on them, they would lead through onto the anchor winch without fouling any of the other gear. The bowsprit had to be able to run back over the barrel of the anchor winch and over the lower barrel of the cargo winch and alongside the mast without fouling any of them. This all works well, last year (1992) we had the bowsprit run out for the Summer and it's a one man job – one man can put the bowsprit out and back again with very little trouble.

To finish off the spars, they were oiled, except the ends – the ends are picked out in a charcoal grey colour paint. This colour scheme we also copied off the painting in Gloucester Folk Museum which, as I've said, I'm sure it's pretty true to detail.

Finishing touches to the hold

Continuing with the finishing off of the boat, we needed to fit the stanchions around the cargo hatch for securing the hatch cloths. These were about 30 inches tall, made of iron bar and there are eight per side and three at either end. The height is a deduction but it is a reasonable one. They had to be as high as possible but not too high that you could not get your leg over, because if you can imagine – with the boat working with a full load of cargo, these hatch cloths would have been all around the cargo hold and the crew would need to be able to get over them to get from one end to the other. Its a height that one can just about straddle. Also any higher and it looks wrong – so much of this is rule of thumb, or by eye. All around the edge of the hatch there are sockets in the sides of the hatch coamings. On the for'd and aft ends of the hatch there are actually staples, large iron staples. Two for each stanchion. The stanchions have a flange 6 inches up from their base to stop them sliding right through the socket. They go up to about an 1-inch diameter round section and coming right up to the top they fork into a crutch which takes a 4-inch diameter wooden pole which is lashed in place. The purpose of the stanchions and rail, is for stretching the hatch cloths up over and down on the inboard side to stop waves slopping in over the cargo hatch and swamping the vessel. To lash the hatch cloths on the inside of the cargo hatch, there are staples and rings all around the perimeter of the hatch. These again we made up of wrought iron and forged and fire welded.

Continuing with the fitting out, the main sheet horse was another major piece of blacksmithing. The main sheet horse is really where the boom is sheeted to and is a traveller across the aft side of the cargo hatch. This traveller has a large wooden block shackled to it, where you have a five fold purchase up into the main boom. It has to take tremendous loads – the boom alone must weigh ¼ ton. By the time you've got a whole bellyful of wind in the sail and the boom is thrashing from one side to the other, particularly when the boat jibes, you're putting several tons of weight onto this fitting. It's made up of 1½-inch iron bar, forged around at either end and bolted and spiked onto the hatch coming – galvanised as all the fittings are. We had to fit stops at either end of the horse before it was fastened in place to stop the traveller getting too far round the curve near the fitting and becoming foul – because if you go about and it becomes foul to windward, then it's a devil's own job to shift it.

The bilge pumps

That completed all the major ironwork – other than the bilge pumps. We have fitted three bilge pumps on board the boat, although we don't know how many there actually were originally. There are two aft and one forward. The idea of having one forward is quite universal in barges – particularly where they're loading and offloading from a beach – when they take the ground, the boat can settle bow down and all the bilge water goes up that end – also if you then spring a leak before the boat floats off again you need to have means to pump it out. So the idea of the pump forward is so you can pump the bilge out if she's down by the head.

Aft, I put two pumps in – one was a primary and one a secondary pump. They are two different sorts, these aft two. The primary pump is a very large diameter, about 9-inch bore with a very short stroke. The other pump is the same as the forward pump and is only about a 6-inch bore and a much longer stroke.

The pump forward and the secondary pump aft, were the ones we salvaged off the Thames barge the *Monarch*, which is wrecked just behind the lock at Tewkesbury. These are very primitive pumps and have no mechanism at all – they have a wooden clacker valve – a housing with a leather valve with a lead weight on top. This is the bottom diaphragm or bottom valve. The plunger is another wooden plug with a leather washer to give you your interference fit to the side of the pump bore. That is pierced by a hole bored out and there's a leather clacker valve on the inside of that which gives you your top valve. The pump just discharges onto the deck and the water washes off the edges of the gun'ls.

Of course we only salvaged the actual iron parts of the pumps – we had to make up the elm bottom plug and the valve, plus the plunger block and valve and then the connection rods, the stanchion which the handle operates from and the handle – these were again forged up and galvanised. The bores of the pumps were very badly corroded and I took the decision that though we could probably get them bored out – although at quite an expense – unless the pumps were used on a regular basis, the bores were always going to rust up and a rusty bore will just shred any form of plunger that we're trying to work up and down it. So, after bonfire night at the Museum, we dropped the bilge pumps into the remains of the fire and built the embers up over them and got them to a dull red colour. This lifted off 90% of the rust. When they cooled, the contraction popped almost all the rust off. The rest of the rust we worked off with long cold chisels, working down the bore. This gave us quite a clear bore of just over 6¼-inch diameter and I was able to buy some copper pipe – quite expensive – about £250 for 5 foot of copper pipe with a 6-inch bore and I think 3⁄16-inch wall thickness. You can understand it's not a normal 'off the peg' commodity and I was lucky to find the supply. But it's made the pumps serviceable and useable for another 100 years plus. This copper sleeve we actually put in with glue to seal it from moisture, so the bore doesn't rust between the copper and the cast iron of the body.

These pumps must weight about 130–140 lbs each so they're quite substantial bits of casting. They were actually made by Gardener & Co.

The other pump is a Loveridge's 'New Deluge' – quite a smart name in fact for this piece of equipment. This is the one that was found on the beach at Lydney with the cast-iron ends in two separate areas and with the barrel missing. We had no details of what this pump was like originally, other than that I knew there would be some sort of barrel between the top and bottom sections. I got a piece of steel pipe machined up as the barrel and had it galvanised. I had it made 9 inches long. Subsequently, I actually found a boat, in London, which had got exactly the same pump on board it, and the actual barrel height was 6 not 9 inches. But at this stage ours was all bolted down on deck. I was loath to disturb it at this stage so it'll have to survive with a 9-inch barrel height. Also ours is of galvanised steel, not bronze, as the original one would have been. However, what I did manage to do, was to copy details of the pump on this boat – it was called the *Kathline and May* – she's tied up in London. I got permission from the Maritime Trust, who owned her, to dismantle the pump, because we had none of the mechanism for ours. I borrowed the castings, brought them back home here and got them recast in bronze and cast iron respectively. These were for all the moving parts, the valves, the plunger, the rocker arm, and these were a great success really. The Museum Iron Foundry cast the cast-iron rocker arm for us. We also moulded up the bronze castings here at the Museum, but we do not cast bronze, so we took the moulds very gently by lorry up to Telford Shell Castings Ltd. at Halesfield and there they poured the bronze for us. That went very well too. There was quite a bit of bubbling from interaction between the bronze and the moulding sand but luckily when we came to machine the castings, all the drossy bits machined out.

The next problem was the actual washer or the part which forms the seal between the barrel and the plunger. It consists of a 1-inch square hemp braid packed with tallow, like packing on steam glands. I spent nearly half a day on the telephone phoning round all different places and eventually traced a Company up in Lancashire, in Rochdale. It was T. B. A. Industrial Products Ltd., actually their Sealing Materials Division. This Company there is still making this greased hemp gland packing and they are probably the only Company in the world still making the stuff, but as the demand these days is so little, they only make very limited runs of it, about once every 5 years. The machinery is about 100 years old they make it on, but because there's so little demand there's no point in upgrading it. Each machine has to be specially set up for making whatever dimension of braid. They said the price was something like £200 for a 5-foot length, which is astronomic, but it was like the copper pipe – we needed it and felt we had to pay for it. Actually they were starting up a run just that week, which was lucky for us. They said we could have the first eight metres off it – and they gave it to us for nothing, which was very, very good. So we got the real McCoy – we got hemp braid made to our specification and we'd got enough left for a replacement if it ever wears out. It's good to have some spare bits, I don't think it will be available at all much longer.

After we had installed these pumps, when we put them to work for the first time, just sucking air. It seemed amazing – the clacker valves jumped around on their seatings and actually sucked air up from below. Which is not bad considering we'd got no water helping to seal the valves. When she, hopefully, actually does go sailing, I'm sure they'll work without any trouble at all.

Fitting out the cabins

We next got on with fitting out the cabins. Paul fitted out the for'd cabin and I fitted out the aft cabin.

Now we had found some evidence, when we dismantled *Spry*, as to how they were fitted out. Under the foredeck there was a row of nail holes on the for'd edge of the anchor winch legs and another row of nails on the outboard aft edge of the anchor winch legs, obviously where there had been a batten or a piece of timber nailed to the legs which carried a bulkhead or a partition or a cupboard. There was also a row of nails along the ceiling of the boat, which gave us, what I believe to be the height of the bunks. The fact that there was no evidence of any paint on the ceiling or the lining of *Spry*'s cabins, when we came dismantle her, seemed to show fairly clearly, that these whole areas were panelled out. We did find traces of paint on the deck beams, and these were coloured blue with traces of white at the edges where obviously the paint had overlapped. I've since found out that blue and white were the colour schemes of Jacob Rice & Co. who were the boats owners in the '40s. They used to paint out the cabins every year, and there was so much paint on the cabins you couldn't open and shut the doors half the time. If they were painted out so regularly, either the sides of the boat weren't painted, or they were panelled. It seems very reasonable to presume they were panelled.

We really based the layout and finish mainly on these clues we found. We knew there were little cupboards forward, outboard of the anchor post. Two bunks, port and starboard, were of the heights and widths indicated by the rows of nail holes. The bit in between the top of the bunk and the underside of the deckhead was panelled in.

We weren't sure whether there were two or four bunks in the forepeak. She's certified to carry four crew but from her records, she only worked with four crew for the first six months of her working career. She then worked with a skipper and three crew. Whether the mate lived down aft or lived up for'd with the crew, we don't know.

Figure 49: For'd crew's cabin, with bunks port and starboard and cast-iron stove midships.

It is probable that there would have been four bunks aboard the boat. However, not having sufficient information, I decided just to put two bunks in, but there is provision and landings to put another bunk above each of the lower bunks if at a future date we find there were definitely four bunks. We fitted out the whole of this cabin with larch framing and Douglas panelling. It's all mortised and tenoned. They're not 'fielded' panels, they're just planed off on their back faces to fit the rebate in the framing. 'Fielded' panels are ones where you've got a chamfer around the panel and you have a vertical step down onto the chamfer from the frame. These weren't fielded panels like that, they're just flat panels. They were painted up with red lead, undercoat, gloss finished, blue and white. We matched the blue paint with the paint flakes we discovered on the deck beams. I believe the blue is actually called Dresden Blue. (Figure 49)

The skipper's cabin is a bit more refined, as befits his position. Again we did this out with larch framing and Douglas panelling. But in here it was all varnished so that the joints had to be absolutely right. The great difficulty with both cabins is that nothing is square – nothing is parallel – the cabins are fitted to use as much of the space as is possible in these highly shaped ends of the vessel, so this created tremendous headaches in deciding where we started a datum, so we could then lay everything off from that point – for rake, or for following the sheer of the deck or for following the rise of the ceiling. It's worked out quite well, I'm pleased to say. The cabin layout consists of a horse-shoe shaped settle around the back end of the boat, reaching round further to port than it does to starboard. (Figure 50) At the starboard end of the settle, there's the skipper's bunk, which is like a cupboard really, it's nearly totally enclosed, just having an opening to get into it from the cabin. To port, you've got the skipper's companionway hatch

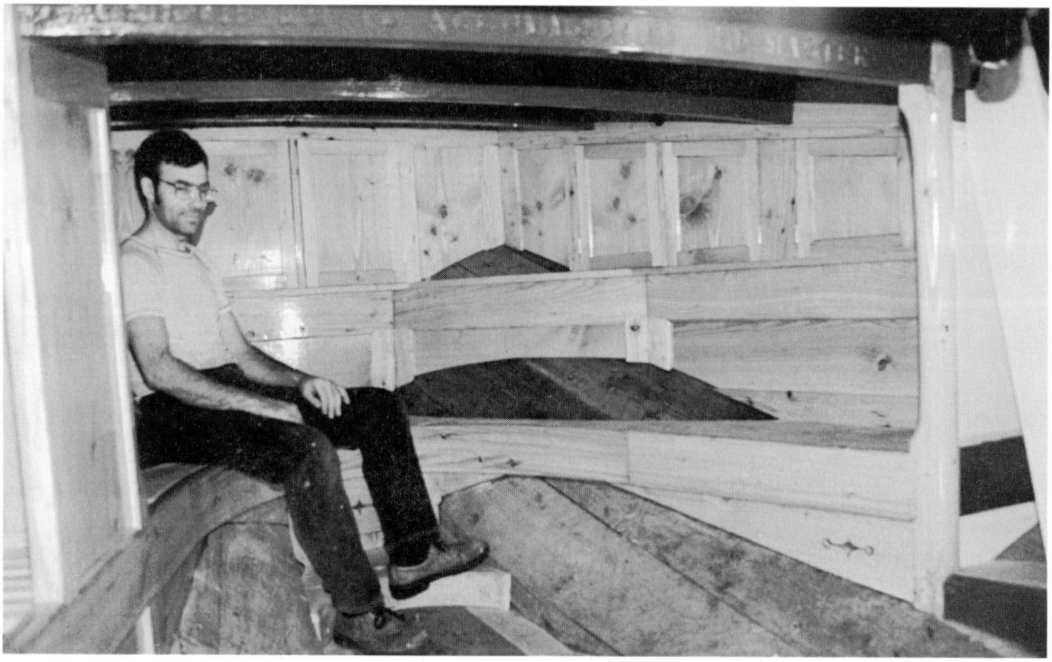

Above: Figure 50: Inside the master's cabin, Paul Turner sitting on the settle.

Left: Figure 51: The wet area at the foot of the master's companionway steps, showing lockers.

with steps coming down and at their bottom there is the wet area where you'd be coming down with wet oilskins and 'wellies' – it's also where the rain drives in if the cabin hatch is open. This area was painted out because varnish doesn't last at all when you get it exposed to the elements. This would have had a cupboard as you came down the steps to port and we have fitted one. (Figure 51) You would turn right into the cabin proper. Behind the settee there's a shelf all the way round and cupboards behind that for storage of all the necessities of life.

The cooking on the boat would have been done predominantly for'd, in the crew accommodation – we had that confirmed the other day by a gentleman who came around to see the boat – at least how it was done in *Spry*'s later life. He was a lad in the 'forties and actually worked with his father on *Spry*. He said that there were stoves in both cabins but meals were usually cooked for'd and then carried aft. I am sure though that the skipper would have had a coffee pot or a kettle on his end of the boat for hour-to-hour necessities!

With regard to the stoves – again, we didn't have any evidence of what they were like, so we've actually bought modern day facsimile stoves. They're multi-fuel stoves, made of cast iron, and about 18 inches square and 20 inches high. They're of a type still used in boats and it's rather appropriate that the motif cast on the side of them is an anchor. To these we've fitted a brass fiddle rail around the top edge to stop pots and pans flying off with movement in rough weather. We also made stove pipes and ash pans. These were mounted on top of cement sheets so if there were hot embers lying around in the ash pan the heat wouldn't scorch through to the timber the other side. The deckhead is also lined with a cement sheet to stop scorching and to keep heat away from the underside of the deck which, if it dried out too much, may start leaking. Definitely not a thing you want happening.

There had to be light in the cabins and we put in light fittings – both artificial and natural. We've got brass lamps hanging in gimbals, there are three in the skipper's cabin and two in the crew cabin. They're of a style that could have been in originally, but we haven't evidence of what they were actually like. They're the nearest thing we could buy today off the shelf. They're paraffin lamps, but for daylight, the skipper has a skylight. This is a raised light of about 2-foot square which stands up about 14 inches from the aft deck. It's got two opening windows – one port, one starboard. Standing on a small platform before the settle, the skipper can get his head up into it so he can keep an eye on the weather and on what's going on. Imagine these boats were often laid up between tides, if the tide's running foul, you'd anchor up and wait for it to change. Waiting for the tide was really when you got your 'shuteye' or your 'kip'. So if the crew's all down below, someone would have to go up on deck to see what the situation is. But the skipper could have quite a good view from his skylight while staying down below. He could also see the set of the sails and so on while under way, from his cabin lookout. The other means of natural light down below is through deck prisms. There's two deck prisms up forward and one down aft in the skipper's cabin. These are just glass prisms, 9 inches long and 3 inches wide, let into the deck and held down in place by bronze rims. They're very effective for getting light down below. We were able to buy these and fit them, they haven't changed over the years and are still used in modern boats. In addition we fitted two portholes, similar to those fitted originally – one in the skipper's hatch companionway, and the other in the crew hatch.

The only things that still remain to be finished in the skipper's cabin are to upholster the settle and to install a table. The settle had an upholstered seat and I've got the leather for that. I couldn't think of any other material that it could reasonably have been made of. If you think of the conditions, they'd be sitting down there in work clothes, often wet and quite heavily soiled perhaps – so you don't want a cloth covering. There weren't plastic coverings available then of course, so I presumed leather would be the natural selection. It would have been stuffed with horsehair or coir and I've actually got 10 lbs of coir stuffing for doing the seat pads, but as yet I haven't got round to making them.

The other thing, the table in the skipper's cabin would have been a free standing table with a loose-leaf or drop-down leaf for when the crew came in to dinner – I believe they all ate together in the skipper's cabin. Also the skipper would perhaps entertain his family or guests or the agents in his cabin – and more importantly he would use it for looking at his navigation charts. I haven't been able to get round yet to deciding how this should be made.

This covers the fitting out of the skipper's cabin – other than the most important matter of providing ventilation. What ruins many boats, particularly when they are panelled out in cabin areas, is lack of ventilation. Not for the skipper or the occupants, but for the boat, for the hull itself, to be able to breath. What I've done which is not original, I've put two mushroom vents in the deck, one up forward underneath the bowsprit so it's not in the way of the working the vessel, and the other one is right down aft, just on the starboard side of the stern post. If you get air down below at least you keep the humidity level somewhat under control and these will let air through. On the inside of the cabins, in the panelling, in these areas I've actually made the vents a decorative feature – like drilling a hole and then cutting it out with a jig-saw to a star shape coming out of the hole. I've backed all these with zinc gauze so no vermin can get through. They look quite effective, actually they look quite Victorian. I've seen it done in cupboards and meat safes and things like that in the past in Victorian furniture, and it looks right in the skipper's cabin.

I enjoyed fitting out the skipper's cabin and thinking of these sorts of important details. It was like having the cream off the top of the pudding doing this, we'd slogged away, nearly busting our guts at times building the boat, to be able to do something like fitting out the skipper's cabin really was a treat – a luxury. Little details like the decorative vents and so on are what catches people's eyes and makes them think twice about the reasons for it. Maybe it will alert people to the problems likely to occur in the future with boat restoration and the importance of ventilation or daylight and so on.

To complete the cabins, the only thing we had to do then was the installation of the companionway hatches and skylights.

The crew hatch is a very low affair – it stands not more than 9 or 10 inches off the deck. It's just a straight sliding hatch. The hatch lid slides aft over runners underneath the cargo winch. Access down into the crews cabin is by very steep companionway steps, and it has a porthole to starboard which is non-opening.

The skipper's companionway hatch is a much more elaborate affair. It's got a curved top and has a curved lid which slides back on curved runners. It is framed and panelled as would befit the situation. It's all made out of pitch pine and has two loose hatch weather boards, aft, which lift out to allow access. It needed all to be made with weatherproof joints, so no matter how much water's chucking about, in theory, not much should get down below. Similar to the crews companionway, there's a porthole to starboard in this companionway, non-opening again. We know there was a porthole originally in the crew hatch companionway, because we've actually got the original hatch for this and the porthole aperture is still there. It wouldn't have been an opening one because you'd have been banging your head on it all the time. For the one into the skipper's cabin, I'd have thought he'd have wanted to be able to look out before actually opening up his hatch. If it's really cutting up rough and you've got a load of cargo on board the boat, there's a lot of water slopping around the deck, or waves breaking alongside the boat, so the last thing you want to do is open the hatch and get a dollop of water on top of you, so I am certain there would have been a porthole there.

The skylight in the skipper's cabin is nearly all glass, it's really just got 4 legs, one in each corner. The for'd and aft windows are fixed windows while the port and starboard ones are opening lights as I mentioned before, so you have a window frame which is glazed and window stays so these lights can be propped open. We fitted a bronze grille on the outside of all four sides because it's very vulnerable to put one's foot through or to be hit by any loose gear rolling around the deck.

I think that just about covers all our work we've done on *Spry*. We think that she is very nearly 100% as she was originally, or perhaps even better in some respects and certainly she's capable of being made fully seaworthy for sailing the estuary again. I hope that it will be possible to continue to finish her completely, to rig her out fully for sailing. In the meantime, although we've made the main mast, as this is far too high to fit in the boat where she is now, we've fixed in a short mast so the the boom and bowsprit can be fitted for temporary display purposes in the 'boat shed' at Blists Hill.

CHAPTER 6

Conclusion

The story of the rebuilding is complete. *Spry* has been faithfully and painstakingly recreated with great attention to authenticity and detail, as the reader will readily appreciate from the previous Chapter. Only two timbers from the former *Spry* as recovered from the Diglis Canal basin have been incorporated into the new structure, non of the others being in good enough condition to ensure a reasonable further life in the vessel. These are the keelson, which in fact was not the original one used in the construction in 1894, but a later replacement, and one oak beam across the open hold, which carried the original carved registered tonnage inscription, which was an original installation, but which had been moved at some time to another position in subsequent repair work.

From the inception of the project it was always the aim to display the restored or rebuilt *Spry* on the River Severn adjacent to the Ironbridge Gorge 'Museum of the River' and adjacent to, or on the site of the old Bower wooden boat Building Yard on the opposite bank. This was felt to be very appropriate as such large estuary trows are known to have been built there. Some of the problems likely to be encountered in displaying the vessel on the river were envisaged but no detailed investigation was carried out to establish the feasibility. It seems that it was this intention which largely influenced the decision to carry out the rebuilding to the high standards required for sailing-worthiness.

During the rebuilding, the further objective was formulated by the USNT Management Board of trying to obtain financial sponsorship for a voyage of the vessel down the river to Bristol, from the present upper navigable limit of Worcester. This to be prior to the final berthing in Ironbridge. The voyage to Bristol, berthing there and carrying out further promotional voyages from there, plus giving other educational opportunities for the study of early river navigation, were felt to be very worthwhile in publicising the Ironbridge Gorge, thus taking full advantage of the high standards of sailing worthiness being achieved. This would have been especially so if the voyage could have been coincident with the centenary of the vessel's original launch date, which is in 1994.

An outline feasibility study of the practical prospects of such a promotional voyage has been carried out and is favourable, but up to the time of writing this account, in the continuing climate of recession, no sponsor yet has been found, but the prospects of sponsorship eventually do not seem unfavourable.

Figure 52: *Spry* complete awaiting eventual sailing preparations, fitted with short mast for mounting boom and spars for temporary display purposes.

With regard to the final display arrangements for the vessel in Ironbridge, it has unfortunately not yet been possible to agree on a clear objective. The original aim of providing a river berth has given rise to differing opinions on the most suitable site and to misgivings regarding security against floods and vandalism and also regarding the not inconsiderable ongoing maintenance costs. For this reason and again also lack of finance, practical feasibility investigation has been limited to contact with the River and local Planning Authorities, who saw no objection in principle to a berth on the River by the Museum of the River. Again at the time of writing, the final display arrangement objective is still to be finalised, whether a river berth or on dry land and the actual siting. This present account cannot therefore be taken to finality, if these ambitions are to be eventually realised, but perhaps at some later date it will be possible to at least add a postscript describing a fitting conclusion to this ambitious and skilful project. The fulfilment of the ambition to make a promotional voyage down River would surely make a fascinating subject for a further book!

In the meantime *Spry* will continue to be housed in the by now familiar 'boat shed' on the lower Blists Hill site of the Ironbridge Museum, almost within sight of its native element, the River, with improved facilities for public access, with a high level gantry to facilitate viewing the decks and inside the hold. A display alongside will give a brief illustrated account of the history of the vessel and its rebuilding and show related artifacts.

Figure 53: Bowsprit run out to sailing position – through the 'boat shed' end wall.

To give the most realistic appearance, in view of the restricted head room in the 'boat shed', a short mast has been made and fitted to carry the boom and the end of the shed has been pierced to enable the bowsprit to be run out – a very incongruous sight from outside! (Figures 52 and 53)

The visitor cannot fail to be impressed by the size, but it is also hoped, using the imagination, with the help of the information provided, that a mental image of *Spry* battling her way across the estuary will be stirred, along with the appreciation that this vessel is the recreation in almost exact detail, of the last survivor of a very long line of similar up-river vessels, which played a vital role in the development of the 'Birthplace of Industry' nearly three centuries ago, in the river Severn Gorge, which became known as the Ironbridge Gorge, following the erection of the bridge which was such an outstanding example of one of its main products. The visitor may also spare a thought for the hardy breed of men who sailed and heaved the vessels up and down the river and those their counterparts on and under the land, wresting its treasures out to be smelted or kiln fired to provide their cargos. It is thanks to these and the vision of the men who directed their activities, that our present sophisticated way of life was made possible.

Last and not least, even the landlubbers amongst the visitors to *Spry*, and these will in all probability be the majority, while not being able to appreciate, or even see much of the detail of the rebuilding, will be impressed by the overall quality and scope of the work which they can see, and recognise the skill and effort of Master Shipwright Alan Williams and his small team of people, particularly those of his very able lieutenant Paul Turner, who has played a prominent part throughout.

APPENDIX I

U.S.N.T. Ltd

Memorandum of Association (extract)

Section. 3. The objectives for which the Trust is established are as follows: To acquire one or more examples of the sailing vessels known as Severn Trows and such other craft as were operated on the Upper Severn and its associated waterways and to equip such craft with sails, gear, cordage, chandlery equipment and apparatus of the sort traditionally used thereon or in connection therewith, and for the education and benefit of the public, to operate, exhibit and preserve such craft in sailing condition to demonstrate their historical significance in the development of water-borne transport on the Upper Severn and its associated waterways, and the contribution made by such craft to the economic and industrial expansion of the area served by such waterways. And in furtherance of those objectives but not otherwise, the Trust shall have the power to do all or any of the following:

(a) to acquire by gift purchase lease hire or otherwise any real or personal property rights or privileges and to sell lease lend exchange or otherwise deal with all or any part thereof in such a manner as may be considered appropriate.

(b) to construct rebuild restore repair extend and improve any waterway building or structure and generally to develop manage and maintain any land and the water buildings or structures thereon.

(c) to make such arrangements as are necessary to enable the public to view and enjoy any exhibit of the Trust (whether free or at a charge)

(d) to undertake or support research into the use of the Upper Severn and its associated waterways for transportation and navigation.

(e) to raise funds by subscriptions, donations, grants, loans or otherwise for the purposes of the Trust and to invite and accept gifts of all sorts whether of money or of property and whether or not subject to conditions and to carry out any obligation attached to any gift which may be accepted.

(f) by publishing books pamphlets or leaflets or in any other appropriate manner to inform the public of the history of navigation on the Upper Severn and its associated waterways and of the development and use of the Severn Trow and any other craft on these waters.

(g) to constitute a special charitable trust for any particular purpose of the Trust or to act as trustee for any such special trust whether constituted by the Trust or otherwise.

(h) to employ staff of all sorts and subject to the provision hereinafter contained to remunerate such staff and pay or make provision for pensions gratuities and like benefits for any who are or who have been so employed or their dependants.

(i) to invest the moneys of the Trust not immediately required for its purposes in or upon such investments securities or property as may be thought fit, subject nevertheless to such conditions and such consents as may for the time being be imposed or required by law, and subject also as hereinafter specified.

(j) to establish and support or assist in the establishment and support of any charitable organisations or institutions, and to subscribe or guarantee money for charitable purposes in any way connected with the purposes of the Trust, or calculated to further its objects.

(k) to cooperate with any local or public authority or other body concerned to achieve the objects of the Trust or any of them.

(l) generally to do all such lawful things as are incidental or conducive to the attainment of the above objects or any of them and do not derogate from the charitable nature of those objects.

PROVIDED THAT:

(i) in case the Trust shall take or hold any property which may be subject to any such trusts, the Trust shall only deal with or invest the same in such manner as allowed by law, having regard to such trusts.

(ii) the Trust shall not support with its funds any object, or endeavour to impose on or procure to be observed by its members or others, any regulation, restriction or condition which if any object of the Trust would make it a Trade Union or which is of a religiously sectarian or a party political nature.

(iii) in case the Trust shall take or hold any property subject to the jurisdiction of the Charity Commissioners for England and Wales the Trust shall not sell, mortgage, charge or lease the same without such authority, approval or consent as may be required by law, and as regards any such property, the Council of Management or Governing Body of the Trust shall be chargeable for any such property that may come into their hands and shall be answerable and accountable for their own acts, receipts, neglects and defaults, and for the due administration of such property in the same manner and to the same extent as they would, as such Council of Management or Governing Body have been if no incorporation had been effected and the incorporation of the Trust shall not diminish or impair any control or authority exercisable by the Chancery Division, or the Charity Commissioners over such Council of Management or Governing Body but they shall, as regards any such property, be subject jointly and separately to such control or authority as if the Trust were not incorporated.

4. The income and property of the Trust whencesoever derived shall be applied solely towards the promotion of the objects of the Trust as set forth in this Memorandum of Association, and no portion thereof shall be paid or transferred directly or indirectly, by way of dividend, bonus or otherwise howsoever by way of profit, to the members of the Trust.

PROVIDED THAT nothing herein shall prevent the payment in good faith of reasonable and proper remuneration to and contributions to pension or superannuation schemes for any officer or servant of the Trust or any member of the Trust in return for services actually rendered to the Trust, nor prevent the payment of interest at a rate not exceeding two per cent per annum over the Bank of England minimum lending rate for the time being on money lent or reasonable and proper rent for premises demised or let by any member to the Trust; but so that no member of the Council of Management or Governing Body of the Trust shall be appointed to any salaried office of the Trust paid by fees, and that no remuneration or other benefit in money or money's worth shall be given by the Trust to any member of such Council or Governing Body, except repayment of out-of-pocket expenses and interest at the rate aforesaid on money lent or reasonable and proper rent for the premises demised or let to the Trust; provided that the provision of the last aforesaid shall not apply to any repayment to any company of which a member of the Council of Management or Governing Body may be a member, and in which such member shall not be bound to account for any share of profits he may receive in respect of such payment.

5. The liability of the members is limited.

6. Every Member of the Trust undertakes to contribute to the assets of the Trust, in the event of the same being wound up while he is a member, or within one year after he ceases to be a member, for payment of the debts, and of the costs, charges and expenses of winding up, and for the adjustment of the rights of the contributories among themselves such amount as may be required not exceeding One Pound.

7. If upon the winding up or dissolution of the Trust there remains, after the satisfaction of all its debts and liabilities, any property whatsoever, the same shall not be paid or distributed among the members of the Trust but shall be given or transferred to some other charitable institution or institutions having objectives which are similar to the objects of the Trust and operating within the designated area of Telford New Town and which shall prohibit the distribution of its or their income and property among its or their members to an extent at least as great as is imposed on the Trust under or by virtue of Clause 4 hereof, such institution or institutions to be determined by the members of the Trust

at or before the time of dissolution, and if so far as effect cannot be given to such provision, then to some other charitable object.

8. True accounts shall be kept of the sum of money received and expended by the Trust and the matters in respect of which such receipts and expenditure take place. of all the sales and purchases of property and goods by the Trust and of the property, credits and liabilities of the Trust, and subject to any reasonable restrictions as to the time and manner of inspecting the same that may be imposed in accordance with the regulations of the Trust for the time being, such accounts shall be open to the inspection of the members. Once at least in every year the accounts of the Trust shall be examined and the correctness of the income and expenditure account and balance sheet ascertained by one or more properly qualified Auditor or Auditors.

APPENDIX 2

Surveyor's Report on *Spry* in Diglis Basin

R. S. Hatchard,
Lock House,
Diglis Basin,
Worcester,
8th November 1981

This is a condition survey for Ironbridge Gorge Museum Trust (Author's note: actually USNT) on the Severn Trow *Spry* 72′ × 18′ 4″ as seen on dry dock at Worcester. *Hammer and prick tested.*

Because of the craft being full of debris and mud, this is a survey on the outside of the boat.

I was asked to give the inside of the boat only a brief inspection.

Originally the boat was sailed but in the latter part of her working life she would have been purely a dumb barge.

At some stage she had been built up about 12″, as the original covering board can be seen now as the lower rubbing strake. This has been cut off flush with the planking on the inside of the boat. It was then, I would imagine, that the stem and stern posts were altered. The original top of the stem post can be seen now as a scarf joint.

The reason for this is obvious – more freeboard and bigger tonnage.

The keel, which has only one scarf in the middle, is elm – 7″ across the bottom. All appears to be sound along the full length of it.

The stem post is oak, but now after various repairs is in four pieces. The middle two pieces are in very poor condition an most of it is loose. The stem bar is there but is hanging off.

The stern post is oak as well. This is in two pieces, the top half having been replaced at some time. The bottom half looks original but contains a lot of rot at the unladen water line.

When the helm was removed, I would assume it was then that all three thimbles to hang on the helm, were burnt off flush with the stern post. The small skeg remains intact, to protect the helm when it is in position.

Bilge rails The boat has a heavy bilge rail either side, approximately 30′ long, 9″ × 9″ elm, designed so that she would sit upright when taking the ground. Also to protect the belly of the boat. These appear to be very sound and although a substantial amount of weight was on them, because of her being out of the water, the fastenings appeared to be in good order.

The transom The whole of the transom planking needs renewing. The bottom planks are soft and at the top on each side, two large holes have been cut in her. Presumably for ventilation, after she was decked over.

At some stage she has had the second and third planks replaced rather crudely in two halves, and the inner stern post has been doubled up each side to accommodate the two halves.

Port side The side and bottom of the boat is in very poor condition. Roughly sixty percent of the bottom has been plated with very thin plating. On removing some of this in places I found underneath, large patches of rot, so soft that you could push your finger through the planking, which I found to be 2⅜″ in thickness. From the water line upwards to the gunwhale is in poor condition, most of the planking is loose. Roughly in every 15′, in each plank I found rot. There are signs of some new planking having been put in latterly in her working life (two planks on the fore shoulder) which are oak, about 40′ in all, but again the fastenings are loose.

The original planking appears to be pitch pine. There are also various small pieces of timber spliced into the planking in different places, where she has had slight and cheap repairs done.

Elm sheathing around the unladen water line, 12″ × ⅜″ elm sheathing, has been nailed on all around the boat, with felt or bedding of horse muck and tar underneath. This has been put on latterly, supposedly to protect bad planking. It is a known spot for craft to decay in.

Starboard side Again this side and bottom of the boat is in a very poor condition. The bottom on this side has been about ninety percent tin plated. The same as on the port side, which on removal of various parts I found large patches of rot, equally as bad as the port side.

This tin plating starts from above the bilge rails and continues right down underneath to the keel, on both sides of the boat. The planks are large, 12″ underneath, coming up the sides they vary from 10″ to 8½″ tapering off each end of the boat.

This side is slightly better than the port side because latterly she has had roughly 86′ of planking replaced in oak, in four planks. These planks range from 25′ to 13′. They start from the fore shoulder, running back to the stern shoulder, which is all easy work compared with each end of the boat, where the planking on both sides, fore and aft, are very soft.

Beams At one time there were five large beams across the boat. Two have been removed, possibly when she was built up, or when she was made a dumb barge. This could have been all at the same time. The beams measure 15½″ × 7½″ (forward) 10″ × 13½″ (stern). They all appeared sound and have good iron hanging, and lodging knees either side, bracing against the beam-shelf and the lining.

All after deck beams have gone, as well as the forward deck beams. None of the deck planking either end can be saved.

Bulwarks The after bulwark is basically all there but a lot of the boarding is missing. On the port side the last two uprights have been cut off flush with the deck. She has two good fareleads on top of the bulwark on either side.

The shackles and eyes still remained either side of the bulwarks for the steering chains, as she would have been wheel steered originally.

The forward bulkhead is more or less intact, apart from some loose boarding. The hawser eyes are still intact either side of the stem and the fareleads on top of the bulwarks.

The winch and main timber head is all there and in fair condition, but both are ready to drop into the bottom of the boat at any time. The same applies to the accommodation hatchway.

I feel that these should be saved before they are damaged as they are basically all intact.

Some of the deck fittings for the movable hand rail can still be seen, either side of the boat.

On briefly inspecting the inside of the boat I found almost every timber and hull plank fastening to be gone. (The fastenings were loose as I have already mentioned and the timbers rotten.)

On the port side latterly she has had the top lining board replaced in pitch pine.

The kelson I found to be 16″ square pitch pine, possibly not the original one, as it was in very good order. The fastenings appeared to be all good.

Having come to the end of this survey, I am afraid I must say, that the *Spry* is just a rotting hulk, quickly returning to nature.

R. S. Hatchard.

APPENDIX 3

Glossary of Terms

Acrow Prop	Proprietary name for a support which can be elongated mechanically, like a jack.
Apron	Broad piece of timber on the inner side of a boats bow, strengthening the stem and picking up the (hood) ends of the planks where they abut the stem.
Batten	Strip of wood used for fastening.
Beamshelf	Timber on which a beam rests.
Bearding	The tapering down of the stem from sheerstrake to keel
Bevel	Chamfered, or sloping edge.
Bilge	The rounding of a ships bottom, or the space inside the bottom of a vessel.
Bowsprit	Spar which projects forward from the bow of a sailing vessel.
Bulkhead	Transverse partition within a ships hull.
Bulwarks	A kind of palisade or fencing surrounding the deck.
Butt	(tree) The lowest part of the tree trunk.
Butt End	(Sawn Timber) Squared off end of a plank or timber.
Cant Frame	Frame in the fore or aft part of a vessel, not in line with the main frames which are square to the keel.
Carvel Planking	External planking where the plank sides abutt, giving a smooth contour.
Cavel	Large wooden cleat fastened to the bulwark stanchions, for tying ropes on.
Cavetto Line	Hollowed (concave) moulding along the hatch side, usually picked out in distinctive paint.
Carlings	Pieces of wood fastened between the deck beams in a fore and aft direction.
Capping Rail	Smooth rail running along the top of the stanchions which form the bulwarks.
Caulking	The filling of the gaps between planks with oakum, to make the hull water-tight.
Ceiling Planking	The inner planking of a vessel above the floor, up the sides.
Chock	Block or piece of timber used for various purposes.

Cleat	A shaped piece of hardwood for fastening a rope to.
Clinker Planking	External planking with the plank edges overlapping downwards.
Coaming	The framing or raised borders around hatchways, skylights etc.
Coke (not fuel)	A square wooden peg inserted into the faces of two pieces of wood joined together, to prevent any sliding movement between them.
Cord Wood	The small branches and twigs from a felled tree. A cord being the measure of quantity in which they are sold.
Cover Board	Plank or strake of planking, placed horizontally, to cover certain parts or openings in a vessel, e.g. the top of the frames, from outer to inner planking.
Cross Tree	Piece of light timber, placed upon the trestle tree or supporting platform of a lower or top mast.
Deadwood	Block of timber used as filling piece or backing support, e.g. as fitted against the inner side of the stem upon the fore end of the keel, to which it is bolted.
Deadeye	A flattened circular piece of wood, perforated for the reception of ropes or lanyards.
Dolly	A heavy metal object bearing against the pre-headed end of a rivet giving a support while the other end is headed by hammering
Dovetail	A kind of dovetail-shaped overlapping joint in wood, used to give extra interlocking strength.
Drip Bead	A projecting beading on the underside of a plank to localise drips of water.
Dumb Barge	Towed vessel with no independent means of propulsion.
Dump	A blunt-ended friction fastening, driven into a smaller diameter pre-drilled hole.
Dutchman	A wooden plug inserted to fill a hole, usually diamond-shaped.
Fairing,	Smoothing the frames or the planks edges by means of a plane, to achieve a clean contour.
Fairlead	A block of wood (or metal) grooved or holed through and fixed, usually either on the bow or stern, but also sometimes elsewhere on a vessel's deck, serving to keep cables, warps etc. in a desired direction.
Fashion	To hand make an object out of wood or metal

Fashion piece	A decorative feature.
Floor Timbers	Pieces of timber forming the floor of the hull frame.
Fid	A short piece of hard wood, usually square in section, mounted in the samson post to lock a retractable bowsprit in the sailing position.
Forefoot	The area of the lower part of the stem adjoining the keel.
Forestay	Wire rope supporting the mast from the trestle tree to the stem.
Futtock	One of several frame timbers extending from the central floor timbers around and up the sides of the hull to the top timber.
Gaff	An upper spar, usually having at one end, a jaw clasping the mast, supporting the top of a main fore and aft sail and often the bottom of a trapezoidal shaped gaff sail.
Garboard	The lower-most strake of outside hull planking, the lower end of which is rabetted into the keel. Often thicker than the rest of the planking.
Gimbal	A contrivance in which a lamp, compass, etc., may be freely suspended so as to maintain a horizontal position independent of the motion of the vessel.
Gunwale (Gun'l)	The upper outer edge of the plank sheer.
Hatch	A rectangular opening in a vessel's deck for access to the inside of the hull for crew or cargo.
Hanging Knee	Bracket in wood, or usually metal, fitted vertically under a beam end to connect it to the vessel's side.
Hawse Hole	One of the holes in the upper bow, near the stem, through which anchor ropes or cables pass.
Hawse Timber	The timber in which the hawse hole is cut.
Hawse Pipe	Pipe, lining the hawse hole to give wear resistance.
Horse	The horizontal iron bar installed laterally in the stern of a vessel to which the main sheet block and traveller are attached.
Huddis	The name of the inventor of a mathematical table for calculating the volume of useable timber from measurement of the lower girth of a tree.
Joggle	Joint of two pieces of timber, contrived to prevent their sliding no one another, with a notch in one piece and a corresponding projection in the other.
Keel	The lowest part of a vessel, which is its 'back-bone'.
Keelson (Kelson)	A substantial backing timber to the keel on the inside of the lower frame, bolted through the frame and the keel

Kedging	Using a light-weight anchor (kedge) to haul a vessel from one position to another, e.g. off the shore.
King Plank	The central plank of a deck, from which the other planks are laid outwards to the sides.
Knighthead	The foremost frame timber in a vessel, one being placed either side of the stem and often rising well above it.
Lee	Sheltered situation in relation to the wind direction.
Limber Hole	Aperture for the passage of water along the length of a vessel through the frame timbers of a vessel's floor.
Lining Planking	Term given to the totality of internal planking and boarding, with which the inside surfaces of a vessel are lined.
Lodging Knee	Bracket in wood, or usually metal, fitted horizontally for connecting beams to a vessel's side, or carlings to beams.
Margin	The line at which the hull planking of a square-stern vessel ends, meeting the near vertical square planking of the stern.
Moulded at . . .	Measurement of depth or thickness.
Moulding	Battens and sculptures of wood (or metal) used for ornamentation.
Nigger head.	The warping drum an the end of a winch.
Pawl	A hinged or loose metal catch used to engage in the teeth of a ratchet or cog wheel, to allow motion in one direction only.
Peen	To spread the plain end of a headed metal pin to grip the components it is to fasten together, after assembling them together.
Pintles	Metal hinge rings, bracketed onto the stern post and on the front edge of the rudder, through which the rudder hinge pin fits.
Rabbet	A step-shaped recess cut along the corner of a timber to accept the edge of another timber.
Ratlings or ratlines	The light ropes attached at intervals to, and crossing the shrouds of the mast rigging, to form steps for the crew to mount aloft.
Rig	The arrangement of masts and rigging of a vessel.
Rigging	The totality of a vessel's masts, spars, etc. with their standing and running ropes.
Ruffle	A corrugated, circular gear wheel, with the ap-

	pearance similar to the gathered lace on cuffs and at the neck of medieval garments.
Samson Post	The strong vertical supporting post, resting on the keelson and extending upwards above the fore-deck and fastened to the deck beams.
Scallop	A series of rounded projections at the edge of a wooden piece used as a form of decoration.
Scantlings	The sectional dimensions of a timber, frame, post plank etc.
Scarf (or scarph) Joint	Joint of two pieces of timber by the overlapping of their shaped ends, the length of the scarf being usually about 9 times the thickness of the connecting pieces.
Scuppers	Apertures serving to lead off small quantities of water overboard, from a vessel's deck.
Scuttle	A small aperture in the deck, in the side planking of companionway etc., principally to admit light and air.
Sheave	The disc of hard wood or metal in a block which revolves with the working of a rope or chain passing over it. Similar to a pulley wheel.
Sheerstrake	The uppermost strake of planking in the topside of a vessel.
Shutting Plank	Plank forced in a gap left between two neighbouring planks, to wedge them aside in order to give tight fitting joints.
Sided at . . .	A dimension. Made to a width of . . .
Side Cloth	Canvas, fitted to rails above the level of a hatch coaming to prevent waves splashing into the hold.
Spile	A form of template for reproducing the dimensions of one timber or plank in making another.
Spike	Long sharp-pointed rod, usually in metal, for fastening timbers together; a form of nail.
Stanchion	A fixed upright support.
Staysail (Stays'l)	Triangular sail carried on a stay, e.g. fore staysail, between mast and stem.
Stealer	A strake of planking which does not run from end to end of the hull.
Steaming	Heating up wooden planks in steam to soften them for easier bending.
Stem	The heavy curved piece of timber, which terminates the fore end of a vessel's hull.
Strake	A continuous line of planks end to end, running all the length of a vessel's hull, from stem to stern.

Stringer	A longitudinal piece of timber, fitted to provide additional strength.
Tabernacle	A long narrow casing forming a well, enclosing and supporting the lower end of a mast, with the aft end being open to allow the mast to be lowered to pass under a bridge.
Tenon	The shaped projection of a piece of timber to be inserted into a corresponding hole or mortice in another timber to make a firm joint.
Thrapping Irons	Large iron staples set into a winch barrel to prevent excessive wear from the three thrapping (wrapping) turns of the anchor chain, rendered around the barrel.
Topsail (Tops'l)	The top-most sail behind a mast.
Tracheid	Water-conducting element in the cellular structure of wood (or any vascular plant).
Transom	The square stern structure of a vessel, or a timber used in the construction of a square stern.
Trestle Tree	Strong pieces of wood, fitted in a fore and aft direction on top of a mast, serving to support a topmast (or crosstrees of a topmast or topgallant mast).
Trunnel	A wooden nail or peg.
Tumbler (rigging)	A block of hard wood to take the thrust of the gaff, irrespective of the angle it is at.
Tumble-home	The inward leaning which may be built into deck stanchions.
Way (in the way of)	Near to.
Whiskers	Thin wire ropes supporting the bowsprit in a lateral direction, connected to either side of the bow.
Wrought Iron	A semi-purified, malleable, forged iron, containing impurities in the form of fine strings of entrained slag, giving directional properties and relatively good corrosion resistance to the natural elements.

APPENDIX 4

Spry Line Drawings

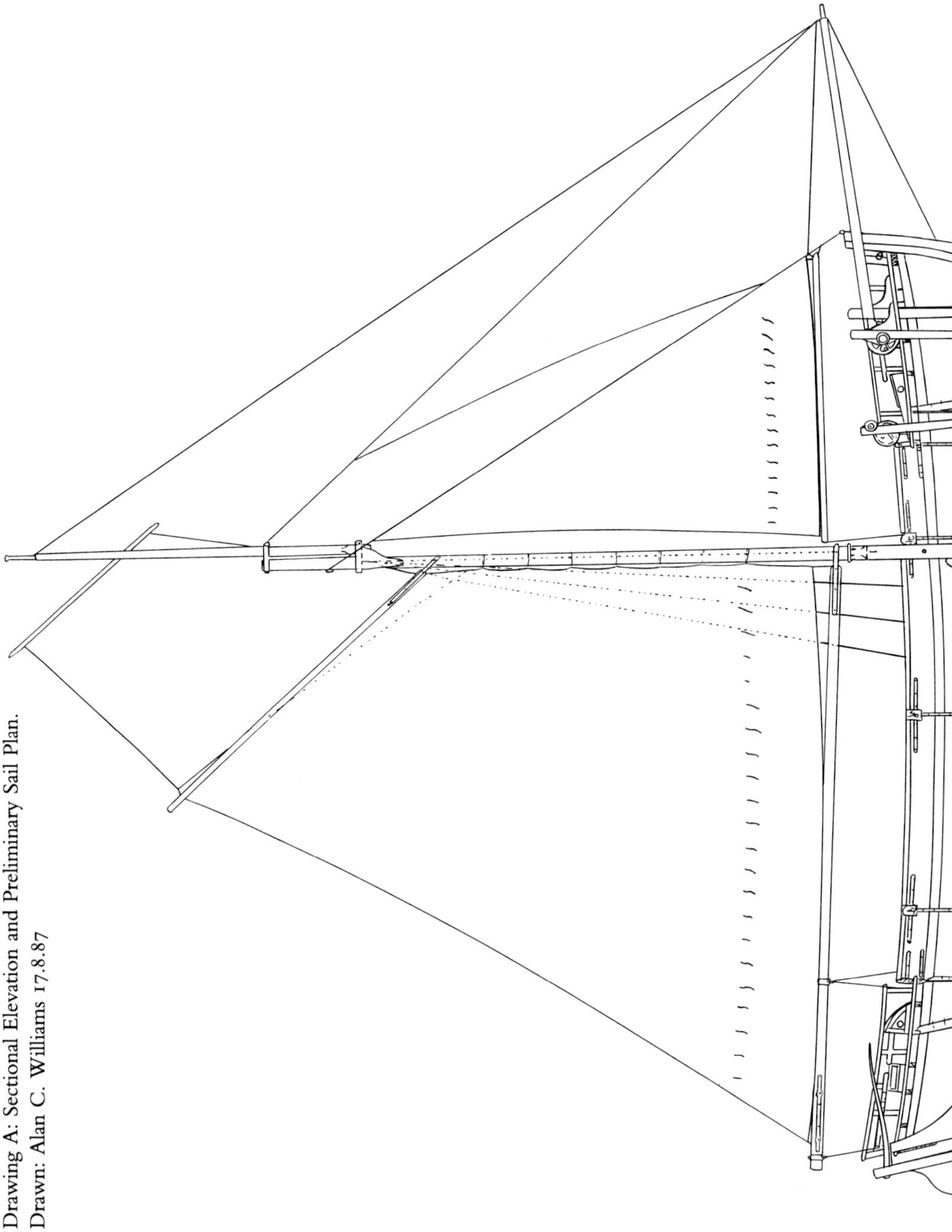

Drawing A: Sectional Elevation and Preliminary Sail Plan.
Drawn: Alan C. Williams 17.8.87

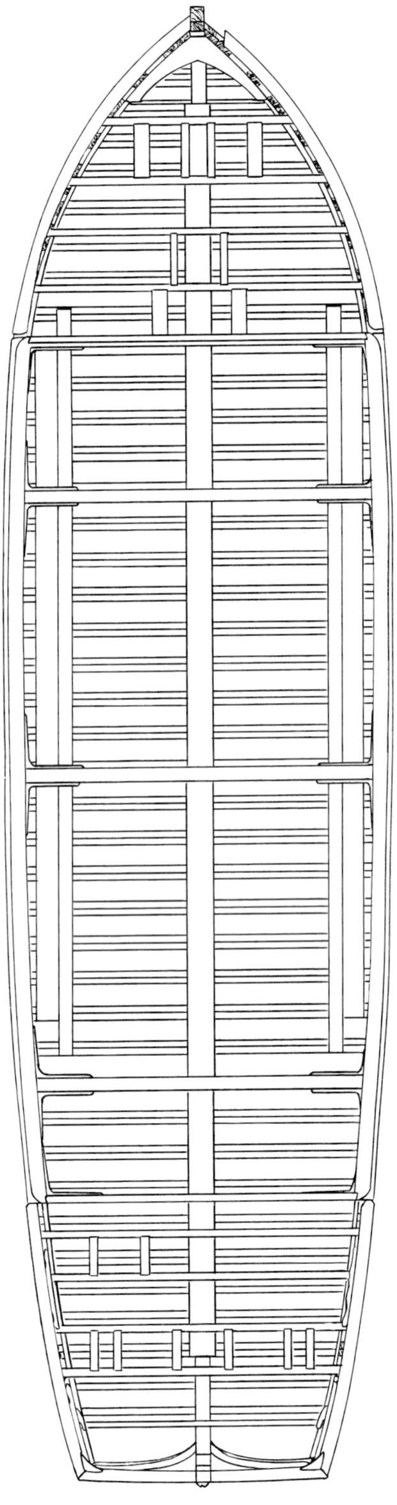

Drawing B: Sectional Elevations and Plan, showing Main Lines and Scantlings.
1) Ceiling left out on plan and side elevations.
2) End elevation sections through A/A.

Drawn: Alan C. Williams 12.8.87

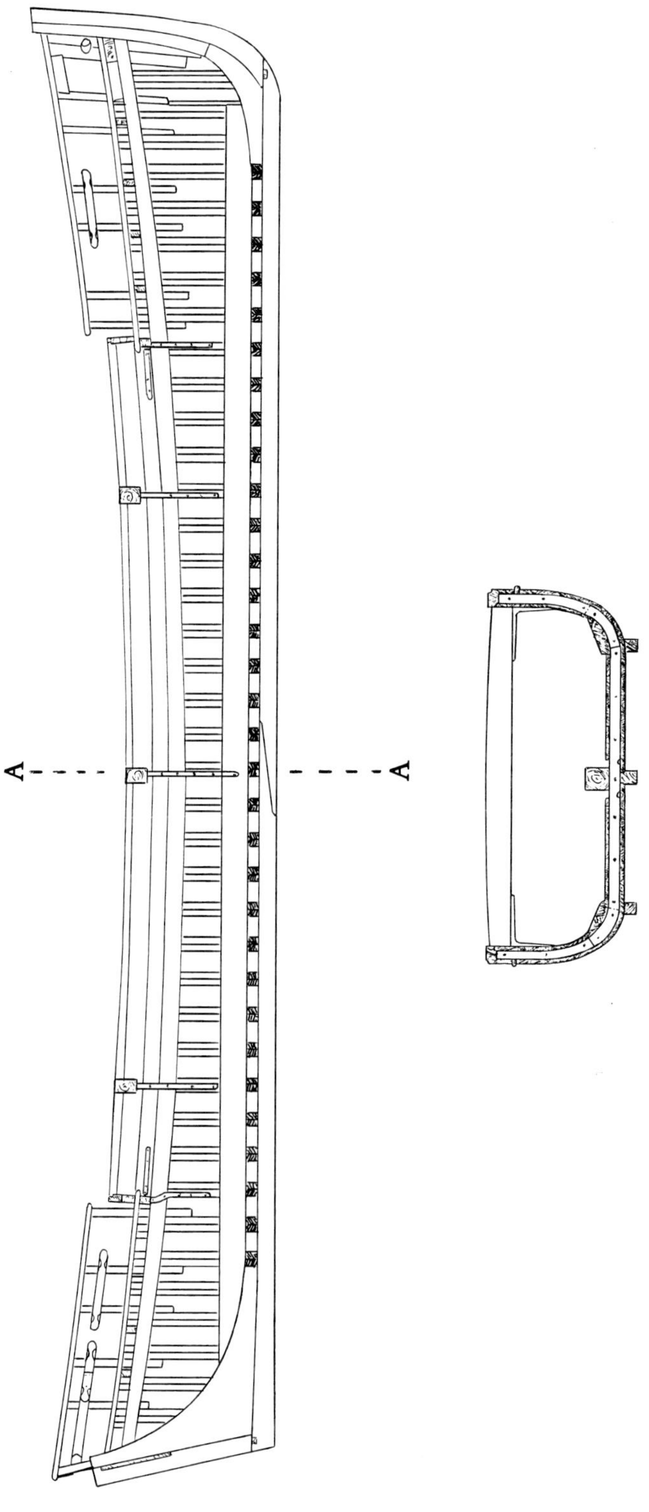

Drawing C: Hull Lines.
1) All lines taken to outside of planking.
2) Hatch sides not shown.
Drawn: Alan C. Williams 12.8.87

APPENDIX 5

Donors to the Spry Project

The Upper Severen Navigation Trust Ltd. is indebted to the following supporters of the project:

Major donors

- Peacock Charitable Trust
- National Heritage Memorial Trust
- Yattenden International Trust Ltd.
- Iliffe Family Charitable Trust
- Ironbridge Gorge Museum Development Trust
- Science Museum
- Kirby Laing Foundation
- Mrs I. Sidaway
- Hedley Foundation Ltd
- Worshipful Company of Shipwrights
- P. F. Holt Charitable Trust

Other national and local benefactors include:

- Birmingham Post and Mail
- C. T. Bowring (Charities Fund) Ltd.
- Peter Bowring Charitable Trust
- Sir Hugh Chance Trust
- Mrs. Ellis and Family
- J. G. Fisher Esq.
- Forestry Commission
- Sir Peter and Lady Gadsden
- Brigadier G. Goulburn DSO DL
- C. S. Johnson Esq.
- Lloyds Bank
- Maritime Museum
- Manifold Trust
- J. B. Morgan Esq.
- T. B. A. Industrial Products Ltd.
- Papachristidis (UK) Ltd.
- M. Rank Charitable Trust
- Shropshire landowners (tree donors)
- Bill Sidaway Memorial Trust
- Sir Adrian Cadbury
- Worshipful Company of Engineers
- Archibald Shaw (Consultancy)
- Swire Trust
- Tanners Wines Ltd.
- Mrs. J. M. Theobald
- Trustees of the *Spry* Preservation Trust
- Transport Trust

Plus many other contributors of useful donations of cash and Gifts in Kind.

APPENDIX 6

Timber Donors

Without the very generous assistance of many local landowners in donating suitable trees, to provide the essential basic materials with which to carry out the rebuilding of *Spry*, the project would almost certainly not have been a feasible proposition.

Special mention therefore is made of the following, who made such donations and facilitated the task of extraction from their land:

The Viscount Boyne, Burwarton, Near Bridgnorth, Shropshire.
The Rt Hon. the Earl of Bradford, Weston Park, Shifnall, Shropshire.
The Hon. Charles Bridgeman, Bowmere Heath, Nr. Shrewsbury, Shropshire.
Lindsay Bury Esq., Munslow, Craven Arms, Shropshire.
The Forestry Commission, Marches Region (Hereford, Shropshire, Worcester).
Mr. Justin Coldwell.
Sir John and Lady Corbett.
Lord and Lady Forester.
Major P. Gifford, Chillington, Brewood, Staffordshire.
The Hon. James Hamilton, Bridgnorth, Shropshire.
Christopher Harley, Esq., Brampton Bryan, Bucknell, Shropshire.
T. Heywood-Lonsdale, Esq., Market Drayton, Shropshire.
Captain J. M. G. Lumsden, Ludlow, Shropshire.
P. L. Marriot, Esq., Powis Castle, Welshpool, Powis.
Col. A. T. Maxwell, Bishops Castle, Shropshire.
The National Trust, Cronkhill House, Shrewsbury, Shropshire.
The Rt Hon. The Earl of Plymouth, Bromfield, Ludlow, Shropshire.
Lord Powys.
Mr. and Mrs. Pugh, Wall under Haywood, Church Stretton, Shropshire.
John Ransford, Esq., Lydbury North, Shropshire.
John Reed, Esq., Sidbury, Bridgnorth, Shropshire.
David Scott, Esq., Cressage, Shropshire.
Nick Stephens, Esq.
Christopher Thompson, Esq., Morville, Bridgnorth, Shropshire.
Sir David Wakeman, Bomere Heath, Nr. Shrewsbury, Shropshire.
Mrs. E. Whittaker.
Simon Windsor-Clive, Esq., Priors Halton, Nr. Ludlow, Shropshire.
Major J. Yates, Codsall Wood, Nr. Wolverhampton, Staffordshire.

Timber Donors by Location

Map locn. no.	Name	Location	Oak	Larch	Douglas
1	Lord Boyne	Burwarton	3	4	
2	Major J. Yates	Codsall Wood	3		
3	Earl of Bradford	Weston-under-Lizard	4		
4	Christopher Thompson	Morville	2		
5	John Reed	Sidbury	2		
6	T. Haywood-Lonsdale	Market Drayton	3		
7	Hon. Charles Bridgeman	Bashchurch	2		
8	Sir David and Lady Wakeman	Rorrington	3		
9	Major Giffard	Brewood		2	
10	Col. Maxwell	Bishops Castle		4	
11	Lindsay Bury	Munsalow		6	
12	Hon. James Hamilton	Bridgnorth		6	
13	Captain Lumsden	Ludlow		6	
14	David Scott	Cressdale		2	
15	P. L. Marriot, Esq.	Welshpool		1	
16	Christopher Harley	Brampton Bryan		2	
17	John Ransford	Ledbury Mortimer/ Priors Hatton		2	
18	Earl of Plymouth	Ludlow			3
19	Tim Sawyer	Forestry Commission		7	
20	Lord Forester	Broseley	2		

Purchases

22	Puch	Church Stretton	2		
23	Anslow	National Trust, Atcham Park	1 Cedar of Lebanon		
24	Barchards	Hull	2		
		Totals	28	41	13
		Plus 1 Cedar of Lebanon			

Copy of letter from Brigadier Goulburn re timber requirements

8 May 1986
From: Brigadier C. Goulburn DSO DL,
Ewdness, Bridgnorth,
Shropshire.

Dear . . . . . .

Operation Spry

The Ironbridge Goerge Museum (author's note: actually the U.S.N.T.), some time ago, recovered the Spry, the last known Severn 'trow' or barge. It is now at Blists Hill awaiting restoration, which will be an expensive operation. With a view to cutting down costs, therefore, I have undertaken to approach Shropshire and adjoining woodland owners, hoping that they will give some of the required timber. In fact I wrote to a few friends as far back as January 1983; and I had many generous results and promises. Work on the boat, however, could not start then because of the problem of finding an experienced shipwright. We have now engaged such a man, Mr. Alan Williams, who has had four years experience building and repairing boats in Australia.

His requirements are about 1,000 cubic feet of oak and 2,000 cubic feet of larch. We want the hedgerow or park type of oak with large heavy limbs from which can be cut the desired curved framing. Williams will explain which of the lower limbs should be left as part of the curve. The minimum length of trunk or butt from ground level to the spreading lower branches should be not less than 13 feet and the diameter at about head height should be not less than 30 inches.

As regards the larch for the planking, we want trees with a minimum diameter of 18 inches at the butt, and 30 to 40 feet in length. They will be a lot more difficult to find than the oak.

It is difficult to convert the required cubic feet into number of trees, at least as far as the oak is concerned, because there will be the inevitable loss or wastage of timber in cutting curved pieces. But, as a rough guess it is probable that 12 oaks and 20 larch will be needed.

The mechanics of the operation would start with a visit by Williams to meet your agent or head forester. He could then perhaps be taken to see the trees you think might serve our purpose, and which you would be prepared to offer. Williams would then say which of these would suit him best.

Big larch is difficult to find in Shropshire; so any offer of such trees would be greatly appreciated.

The Museum will pay for the felling, extraction and haulage, which if done by a timber merchant will be very costly. I am investigating, therefore, the possibility of buying or hiring a mobile band-saw, or a double chain-saw, which will cut the trees in situ where they are lying after felling. The planks and pieces so cut could then be transported in the Museum's own lorry. If the woodland owner had his own sawmill, it would be a great help if he could saw the planks for us.

Can I secure your interest and help in this project? Perhaps you would contact me, either at my home or care of the Ironbridge Gorge Museum Development Trust (Director Major M. G. Harwood-Little, or Mrs. Margaret Prew).

For you information I have sent a copy of this letter to about 40 woodland owners.

Yours,

References

Trow and River Severn Navigation

(For further reading on historical background)

Books in the Ironbridge Gorge Museum Library, in the Long Warehouse, Coalbrookdale:

River Navigation and Trade of the Severn Valley, 1600–1750, T. S. Williams (November 1937). Library Reference 627.1

Severn Navigation and the Trows (extract from the *Mariners Mirror*), G. Farr, April 1946. Library Reference 627.1

Working Boats of Britain, Eric McKee. Library Reference 629.124.